A Touchstone Energy® Cooperative

From its small beginnings in Douglas County, originally formed by a group of locals as the Farmers Electrical Association, GreyStone Power Corporation has grown to become one of the largest, most respected electric cooperatives in Georgia. The year 2011 marks the co-op's seventy-fifth anniversary, and we are proud to share this historic milestone with our members and the communities we serve. The testimonies of powering rural America are perhaps the most singular and stunning reminder of the cooperative achievement itself. *Lighting the Way* celebrates in words and photos the generations and powerful legacy of GreyStone Power. As we honor our past, celebrate the present, and define our vision for the future, we warmly invite you to join us in our journey through this treasured keepsake.

THE
DONNING COMPANY
PUBLISHERS

Copyright © 2011 by GreyStone Power Corporation

All rights reserved, including the right to reproduce this work in any form whatsoever without permission in writing from the publisher, except for brief passages in connection with a review. For information, please write:

The Donning Company Publishers
184 Business Park Drive, Suite 206
Virginia Beach, VA 23462

Steve Mull, General Manager
Barbara Buchanan, Office Manager
Pamela Koch and Wendy Nelson, Editors
Tonya Washam, Graphic Designer
Priscilla Odango, Imaging Artist
Lori Kennedy, Project Research Coordinator
Tonya Washam, Marketing Specialist
Pamela Engelhard, Marketing Advisor

Lynn Walton, Project Director

Library of Congress Cataloging-in-Publication Data

Howell, Kizzy, 1984-
  Lighting the way : a history of GreyStone Power Corporation / by Kizzy Howell.
    p. cm.
  Includes bibliographical references and index.
  ISBN 978-1-57864-704-0 (hard cover : alk. paper)
 1. GreyStone Power Corporation (Georgia)—History. 2. Electric utilities—Georgia—History. 3. Interconnected electric utility systems—Georgia—History. 4. Rural electrification—Georgia—History. I. Title.
  HD9685.U7G734 2011
  334'.6813337932097582—dc23
                    2011024867

Printed in the United States of America at Walsworth Publishing Company

lighting the way

A History of GreyStone Power Corporation, 1936–2011

To the members of GreyStone Power Corporation
1936 to 2011

by Kizzy Howell

dedication

This book is dedicated to the pioneers of GreyStone Power Corporation who made their hope for electricity in rural America a reality, and to the eighty-three courageous members who began the cooperative in 1936. The book is also dedicated to the board of directors and employees who contributed to GreyStone's success throughout the years, and especially to those who gave their lives in service to GreyStone members: Harvey Clark, Travis Allen, and Tracy Worthan.

Perhaps the most important fact about electric membership cooperatives (EMCs) is that they are owned by the people they serve, and at GreyStone, members matter. It's been a pleasure serving our membership these past seventy-five years. We look forward to working with our members and for our members in the future.

table of contents

6	Foreword
7	Acknowledgments
9	Introduction
10	Chapter 1: 1930s: A True Grassroots Effort
18	Chapter 2: 1940s: Meeting the Demands for Service
24	Chapter 3: 1950s: Pushing Ahead
32	Chapter 4: 1960s: Continued Growth
36	Chapter 5: 1970s: Overcoming Challenges
42	Chapter 6: 1980s: Reaching New Heights
48	Chapter 7: 1990s: Preparing for the Future
58	Chapter 8: 2000s: Embracing a New Millennium
80	Chapter 9: Working Toward a Bright Future
90	Chapter 10: Co-op Leaders
96	About the Author

foreword

Although most of the early pioneers of GreyStone Power Corporation are gone today, the light they helped bring to their rural areas still shines through the spirit of their efforts.

As a cooperative owned by the people it serves, GreyStone Power has grown from providing reliable, affordable electricity to offering ancillary services through Gas South, GEMC Federal Credit Union, and EMC Security. From solutions for keeping costs down to helping with energy efficiency, today the co-op is finding new ways to save energy and money by working together inside GreyStone Power and outside with the people we serve.

GreyStone and our members have seen many changes since our inception in 1936, and there will undoubtedly be more as we move into the future. Yet despite what lies ahead, we remain confident in our ability to surmount any challenges with the same resolve as in years past, keeping members at the forefront of GreyStone Power.

Working with our member-owners, we can conquer any obstacle. The philosophy of GreyStone has and will always be to keep the interest of the membership as our number-one priority. This philosophy has served the cooperative and its members well for seventy-five years.

Regards,

Gary A. Miller
GreyStone Power Corporation President/CEO

acknowledgments

The research for and writing of GreyStone's *Lighting the Way* history book was a collective effort that involved a wide network of individuals. Thank you to the people who helped write and publish the book, and to those who helped with the editing and revision of the text. Thanks to excellent recordkeeping, much of the material used in this book was composed through past historical writings, annual reports, and other GreyStone publications. To name everyone who has helped make this book possible without overlooking someone would be impossible. GreyStone especially thanks its members, retirees, and employees who shared their memories and photographs of rural electrification and of the cooperative.

introduction

The history of GreyStone Power Corporation dates back to an era when nearly 90 percent of rural people lived and worked without electricity. In 1936, a handful of west Georgia farmers, homemakers, and small business people worked together to obtain a service that they could not afford individually. The result of their collective efforts rendered a powerful change in their world. Going house to house and neighbor to neighbor, they earned the faith of their lamp-burning public with a newfangled, untouchable commodity known as "the electric."

Like all nonprofit electric membership cooperatives (EMCs), GreyStone Power Corporation operates for the benefit of its members. GreyStone Power is not in the business of distributing electric service to make a profit but only to provide members with reliable electric service at the lowest possible cost. But that's been the idea of the cooperative since its inception in 1936.

Today GreyStone encompasses more than 6,552 miles of lines and over 114,000 meters in its eight-county service area. The cooperative provides electricity to over 103,500 homes, businesses, schools, and industries.

As many of the farms have disappeared from the once rugged service area, residential and business growth have made their mark, urging the cooperative ever forward into a bright future.

chapter one

1930s: A True Grassroots Effort

The history of rural electrification and the tradition of providing affordable electric service began in the darkness of despair, but with perseverance gained success through a ray of hope. In 1882, Thomas Edison built the first central station electric system in Manhattan. Soon afterward, investor-owned electric utilities began providing electric service to cities. Yet, nearly 90 percent of rural people lived and worked without electricity in the 1930s. Investor-owned power companies avoided rural areas that did not ensure immediate profit. Because of their remote locations and rural citizens' low wages, electricity in rural America seemed unlikely. By 1935, only 10.9 percent of local farms had electricity. In fact, many rural people who didn't live close to existing power lines had to do without electricity.

Local residents gathered at farms throughout the service area to learn about this newfangled concept called "the electric" and what it could do for their homes and farms.

Initiated by government intervention, hope for rural electrification became a reality through the Rural Electrification Administration (REA) created by President Franklin D. Roosevelt in 1935. Entering his term in the midst of the Great Depression, President Roosevelt faced a nation of unemployment, foreclosures, and bank failures. He immediately worked to provide relief, recovery, and reform for Americans through his promised New Deal. Roosevelt's administration brought hope to America, touching lives of citizens from all walks of life—even poor farmers living in powerless lands. Roosevelt said, "The test of our progress is not whether we add more to the abundance of those who have much; it is whether we provide enough for those who have too little." His dedication to progress and the establishment of the REA helped to bring affordable electricity to rural areas and made it more feasible for communities to obtain electric service. Slowly, but surely, power lines began to sprout along country roads. The creation of the REA set the wheels in motion that led to the founding of GreyStone Power Corporation.

Although it may be difficult to comprehend today, most of the area currently served by GreyStone Power had no electric service seventy-five years ago. The bright lights of Atlanta could be seen glistening in the darkness from Douglas County and other surrounding areas. However, the lights could have been galaxies away instead of twenty short miles. Georgia Power Company estimated that the cost of construction of power lines in rural areas would be $1,500 per mile. There was an average of only three consumers per mile in rural areas as compared to seventy per mile in the city. The power company did not believe it could turn a profit for stockholders on its investment if the lines were run in remote locations. This, however, was little consolation to the rural residents who for many years suffered the inconvenience and hardships of no electricity. Without electricity, life on the farm was hard. Many of today's farm chores such as milking cows and pumping water were even more time-consuming and cumbersome.

Manager Josiah Abercrombie talks with local farmers and others about signing up to have electricity run at their homes and farms. It was hard to come up with the $5 membership fee.

Chapter One   11

Local people meet, greet, and talk about what electricity can do for them.

As part of President Roosevelt's New Deal program, the REA was the avenue rural Americans had at their disposal to turn on the lights across the countryside and enjoy the same lifestyle their city neighbors had.

On August 24, 1936, five men met at the Douglas County Courthouse in Douglasville and formed the Farmers Electrical Association, a nonprofit cooperative. The original founders, J. H. Abercrombie, J. S. Bomar, W. G. Johnson, N. P. Barker, and R. O. Boatright, together with H. V. Branan and A. A. Fowler Sr. were the first board members. The seven men met on August 29, 1936, to establish bylaws, choose officers, and apply for an REA loan to build eighty-three miles of power lines.

The cooperative's original service area included portions of Douglas, Carroll, Cobb, and Paulding Counties. The seven board members promptly set out on the difficult task of signing up new members for the Farmers Electrical Association. A hard-to-come-by five-dollar membership fee was required to join. "Many people were afraid to sign up because they feared their farms might be lost if the co-op went broke," said Director H. V. Branan. "We had to get at least three homes per mile before the REA would approve the loan. Another problem we faced was the bad roads. Back then, none of the roads were paved in the county except US-78/Bankhead Highway, and it was hard to get around to everybody."

Al Fowler Jr. (Sonny), son of original board member Al Fowler Sr., echoed Branan's view of the difficulty in signing up new members. "The concept of the cooperative went against every known philosophy of power distribution at the time," Fowler remembered. "It was a great debate, even in our own community, whether it could succeed. My father said many people were satisfied with their gas lamps, and then there were the ones who were afraid their farms might be foreclosed by their lenders if the cooperative went bankrupt. But as we all know, the EMC was a great success from the first time they threw on the switches."

When the Rural Electrification Act passed in 1935, the rural electrification movement began all over the country through the actions of men and women, farmers and local citizens, who started a grassroots effort. Member M. C. Austin of Douglasville recalls his memory of the coming of electricity. "At that time we lived in Powder Springs where my dad, Emmett Henry Austin, had worked as a sharecropper for seven years. After giving half of what he made to someone else, he decided there had to be a better way of making a living," said Austin. "He borrowed money from my grandpa, my mother's dad, and took a short course in electricity. In the summer of 1936, he wired house after house in the country." As a boy of eleven years old, Austin helped his father wire homes that summer, going up into the lofts, stretching BX cable through the tongue-in-groove planks. "There was a lot of sweating under those tin roofs," he remembers. The electrical lines Austin and his father installed could carry 110 volts. When the voltage capacity was upped to 220, Austin and his father went back and rewired some of the same houses they wired before so the families could have electric stoves instead of wood stoves.

As a boy, M. C. Austin helped his dad wire homes for the fledgling cooperative.

By September 30, 1936, the Farmers Electrical Association secured an $83,000 loan from the REA. With the loan assured, the Association signed a contract with J. B. McCrary Engineering Corporation of Atlanta to construct eighty-three miles of lines. This would bring central station electrical service to 320 members that the board had signed up through hard work in the four-county service area. In December of 1936, board president J. H. Abercrombie contracted with Georgia Power Company to supply electrical power to the Association.

On June 30, 1937, the board of directors approved the contract between Georgia Power Company and the Association for furnishing energy wholesale. A schedule of rates for consumers was submitted to the REA and agreed upon for the following:

Residential Service—Regular Rate
Residential Service—Optional Rate
Controlled Water Heater Service
Rural Power Service
Commercial Service—Single Phase

The first rate schedule was as follows:
First 25 kWh or less per month	$2.00 per month
Next 25 kWh per month	5.0 cents per kWh
Next 50 kWh per month	3.0 cents per kWh
Next 100 kWh per month	2.0 cents per kWh
Over 200 kWh per month	1.5 cents per kWh

The minimum monthly charge under the above rate was two dollars. Under this rate, the average bill was much higher than today's electric rate when adjusted for inflation. Also, on June 30, 1937, the board of directors officially changed the name of the Farmers Electrical Association to Douglas County Electric Membership Corporation (EMC). At this time, the corporation also became subject to all provisions and restrictions under the Electric Membership Corporation Act.

By September 29, 1937, the fruits of the labor of so many men and women were finally realized. On that day, the first rural line on the

Homemade ice cream lured residents out to hear about joining the cooperative.

cooperative's system was energized, turning a long sought-after dream into a reality. With the successful first phase of the EMC complete, rapid growth was now inevitable. Later in the year, the EMC's board voted to add thirty-five miles of line to the existing system. These new lines would bring electricity to another 114 members. Rural conditions immediately began to improve for the farmers and other residents on the cooperative's system. Douglas County EMC was now providing light to otherwise darkened areas. The lives of rural people and rural society had been transformed forever.

Zula Aderhold Eason, age ninety-one, of Douglasville, a longtime member of the cooperative, reminisced about the excitement she felt when the lights came on.

"Farming was our main source for survival. It was hard work, but we were a happy family. Our money mostly came from selling bales of cotton every fall. We grew our own gardens of corn, fruits, and syrup-cane for syrup. We had no well pumps, bathrooms, washer machines or refrigerator, only an icebox," said Eason.

"I was sixteen years old when the electricity came on in 1936. We were so excited! We already had our home, barns, and chicken houses wired and legally tested and approved. I was in our range chicken house pouring a bag of dry feed into the long tray when the electricity first came on. It was about 1 p.m. I screamed 'whoopee' to the top of my voice! I scared the chickens and they flew into the air and scared me. Then I just stood there in awe looking at our electric lights for the first time. I ran to the house to tell mama about us having electricity. She was happy too! My dad, John M. Aderhold, was one of the first directors for the cooperative. During the planning for the EMC, and many years afterwards, early board members and their families enjoyed planning parties to improve our communities."

Zula Aderhold Eason, age ninety-one

A winter snowstorm catches REA linemen by surprise.

One of the earliest photos of the new electric cooperative's truck after the name was changed to Douglas County Electric Membership Corporation in 1937.

Many residents who were skeptical of the EMC's ability to provide electricity to the area now began to take notice. The once hard-to-come-by membership fees and right-of-way easements were now easier to obtain. In January 1938, the board of the EMC decided to obtain office space for the corporation, having held previous meetings in the Douglas County Courthouse. Office space was rented from Douglasville attorney Astor Merritt, who was also the EMC's first attorney.

Little by little the EMC was spreading its wings and adding equipment and personnel. The president of the cooperative was authorized to employ an engineer to make the final inspection and inventory of all new lines. A half-ton pickup truck was also purchased for the lone maintenance man, and the board allotted funds to build

All over the country, groups gathered around a bare bulb hanging from the ceiling...and the world was changed forever.

a garage to shelter the new truck at the engineer's residence. At this time, there were 118 miles of energized lines on the total system. But with a quickening pace, the number of miles of lines and new members kept expanding.

In May of 1938, the board decided to construct thirteen additional miles of lines. Later that year portions of Fulton and Coweta Counties became part of Douglas County EMC's service area, bringing with them the need to construct an additional 177 miles of lines. In less than two years, Douglas County EMC had more than doubled its entire system.

The first logo after Farmers Electric Association members changed their name to Douglas County EMC.

Thelma Rooks Pate, 98, of Whitesburg

I have lived in the Whitesburg area all my life except during World War II when my husband, Jack Pate, and I lived in Austell and worked at Bell Bomber Aircraft in Marietta. While living there, we enjoyed electricity for the first time. After the war ended, we came back to our home on Old Five Notch Road east of Whitesburg in Carroll County—back home to the dark since we had no electricity there.

About 1946 we traveled in our Model A Ford to visit Dr. W. H. Tanner in south Fulton County who was on the Douglas County EMC board. We talked to him about still not having electricity out in the Carroll County area and how we had tried to convince Carroll EMC to run lines to us. He gave us Douglas County EMC forms to fill out and told us to ask all our neighbors who wanted electricity to fill them out. We drove all over our neighborhood explaining how we had a chance to get electricity out here. Then we took the forms back to Dr. Tanner and he turned them in for us. We paid a five-dollar deposit to get on the lines.

We had our house wired for electricity. At first we just had lightbulbs on the end of long drop cords with a switch on the cord to turn the bulb on and off. The first appliance we bought was a wringer washing machine. Next we bought a stove, and our next purchase was a well pump. We still had an icebox and the iceman brought ice.

Can you imagine what a difference electricity made in our home? We no longer lived in the dark with light only from kerosene lamps. We no longer had to wash clothes in a washtub with a rub board. Cooking over a hot wood stove was replaced by an electric stove, and with our new well pump installed, we had running water. Eventually, we bought a refrigerator too.

We have always been grateful to Dr. Tanner and to Douglas County EMC, now GreyStone Power, for bringing electricity to us.

chapter two

1940s: Meeting the Demands for Service

The first five years of the cooperative saw tremendous growth and showed no evidence of halting. Naturally, as the corporation grew, new positions opened up. Typewriters, adding machines, filing cabinets, and other pieces of office equipment were purchased in order to meet the needs of the growing business.

Every year saw the continued addition of new members and more miles of line. By 1940, the EMC had more than one thousand members. That same year, bids were let for a new office building to house the fast-growing cooperative.

A site was decided upon in June of 1940, and plans for the Douglas County EMC office building were to be submitted in the near future. Local clubs and townspeople helped raise money for the building lot, and the go-ahead signal was given for a new building. On August 19, 1940, a meeting was held to express gratitude to the town of Douglasville and individual donors for donating the lot for the new building for Douglas County EMC.

First real home and first employees of the new electric cooperative off Broad Street in Douglasville.

On November 16, 1940, plans for the building were discussed, and board president H. H. Cook, and vice president Dr. W. H. Tanner decided to choose an architect and procure bids. On November 22, 1940, J. J. Chase, an Atlanta architect, was employed and signed a contract for the building.

In January of 1941, the name *Brighter Homes* was chosen for the monthly news sheet. During this time the co-op was active in repaying as much of its loans as

Employees work hard to keep up with the cooperative growing more quickly than services could be installed.

possible and in getting new loans approved as growth increased and needs became greater. At the Annual Meeting on February 5, 1941, reports were read, resolutions adopted, and directors chosen. By now Douglas County EMC was engaged in the operation of distribution lines in the counties of Carroll, Cobb, Coweta, Douglas, Fayette, Fulton, and Paulding. A movie, *The Power and the Land*, was shown free of charge so members could learn more about the hard work involved in bringing electricity to rural areas.

The EMC consisted of more than 2,200 members and operated 520 miles of lines by October of 1941. Since its inception, the cooperative had obtained its power from Georgia Power Company. With the rapid flight to the rural areas following World War II caused by the housing shortages in cities, it was difficult for the cooperative to meet all of the demands for its service. Georgia Power Company, which only a decade before would not serve areas in the counties where the EMC was located because of a perceived lack of profit to be made, now was battling the EMC for customers. The EMC was at another disadvantage because of the shortage of material needed to keep up with the huge demand.

In June 1942 at a special meeting, the directors voted for Douglas County EMC to become a member of the National Rural Electric Cooperative Association. The

The first inside staff of Douglas County EMC. (Front row, l-r) Mary Alice Mayfield, Annice Hallman, Estelle Abercrombie, and Mildred Dukes. (Back row, l-r) J. H. Abercrombie, Huey Gibson, Theckle Scofield, and Rachel Holloway.

Chapter Two 19

Operations personnel who brought the co-op to life.

Mid-1940s outside employees (l-r) Harold Landrum, Don Giles, Robert Hunter, Clyde Rice, Wilbur Hensley, H.T. (Red) Mason, Howard Teal, Roy Lawler, Elmer Hudson, and Julian Vance.

directors voted to change the date of the Annual Meeting to the first Wednesday in August, beginning in 1943. Because of the war, these years were filled with trials and hardships, deprivations and austerity. Materials were scarce or entirely unavailable, but Douglas County EMC struggled forward. During this period material for serving members was obtained by means of credits for minimal units, an absolute necessity for extensions. By November 1943, this complex business of directors, officers, members, wholesale power, new construction,

Laying cable was no easy task with minimum supplies and tools available, but the linemen got the job done.

Douglas County EMC General Manager J. H. Abercrombie addresses an early Annual Meeting at Douglas County High School.

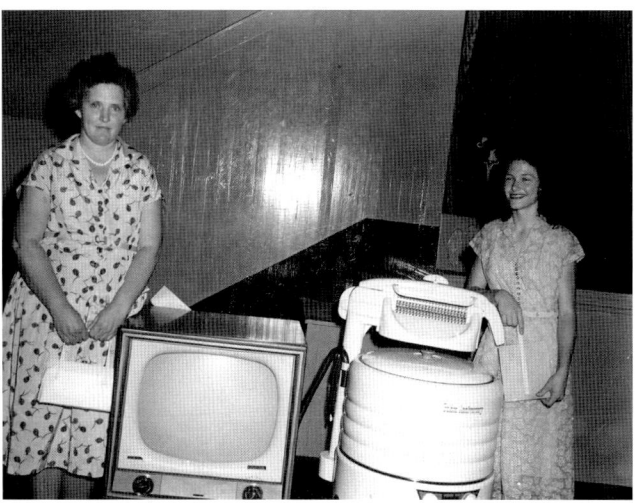
Electrical appliances given away as door prizes brought many to the Annual Meeting.

loans, easements, and service to all areas was nearly as difficult as its first five years of history. Construction was slower because of material shortage, but with the determination and fortitude of pioneers, Douglas County EMC put down tenacious roots and grew steadily stronger. Service was increased, rates decreased, and applications for membership multiplied.

To help rural electric systems achieve their objective of complete area coverage in the territories where they could serve, Congress passed the Pace Act in 1944. This amended the Rural Electrification Act by extending the repayment period for REA

A very early annual meeting drew a crowd; a wringer washing machine was the grand prize.

Chapter Two 21

Willie Wiredhand became a well-known symbol of rural electrification, as shown on Lineman Loyd Lee's shirt, donated to GreyStone.

A mid-1940s EMC board meeting includes (l-r) J. H. Abercrombie, E. T. Evans, W. A. Foster Jr., R. O. Boatright, H. H. Jones, J. R. Mann, Dr. W. H. Tanner, C. G. Rakestaw, and H. H. Cook.

loans from twenty-five to thirty-five years and by setting the interest rate at 2 percent. The floor debate in Congress at that time made it clear that the intent of the legislation was to electrify all of rural America. About 43 percent of the nation's farms had electric service by then.

Lucky winners took home everything from a tire to a refrigerator.

In May 1945 the board voted to subscribe to the newspaper *Rural Georgia* for a one-year period as a trial, with the mailing list to be the permanent users, or the co-op members. By 1947, the cooperative had grown to such an extent that the board of directors decided to sell all of its holdings in Coweta County. This transaction marked the beginnings of what is known today as Coweta-Fayette EMC. At the end of the year, Douglas County EMC was fast approaching 5,000 members and had 775 miles of lines operating. That year, to ensure proper representation of the board of directors, the board voted to divide the cooperative's territory into nine separate districts. The bylaws were amended to provide that one director would have to reside in each one of the nine districts, giving members of each district democratic representation on the board of directors.

By 1949, the EMC had already outgrown its original office space. A loan of $35,000 was secured from the REA to construct additional office space. The cooperative now had six trucks, two pole trailers, and a business car. By June 1949, a two-way radio system was installed, saving mileage on service trucks and time responding to outages and trouble calls. A new substation was also built near Palmetto.

Loyd Lee, GreyStone Retiree

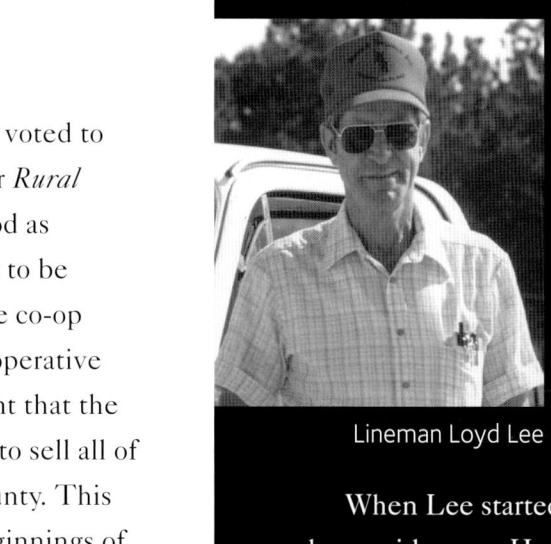

Lineman Loyd Lee

Loyd Lee came to work for Douglas County EMC in 1947 as a groundman, which he says involved "digging holes and lugging poles." Lee worked his way up at the EMC. He learned how to climb a service pole and later became a lineman, a position he held for several years. He was a staking engineer for ten or twelve years and also held the title of branch service supervisor at the Dallas office.

When Lee started at GreyStone, fifteen men composed the outside crew. He says when it came to working on a seven- or eight-pole job, a crew would spend one day on right-of-way, another day on digging holes and setting poles, and a third day on stringing the wires. "Kids would see us coming and say, 'Here comes the REA.' They knew they were getting power," said Lee.

In a time when there were no bucket trucks, Lee sometimes endured very cold temperatures on the job. He traveled with his crew on the back of a flatbed truck with little protection from the wind. "We stopped at a little country store to warm our feet before we started the job," Lee remembers.

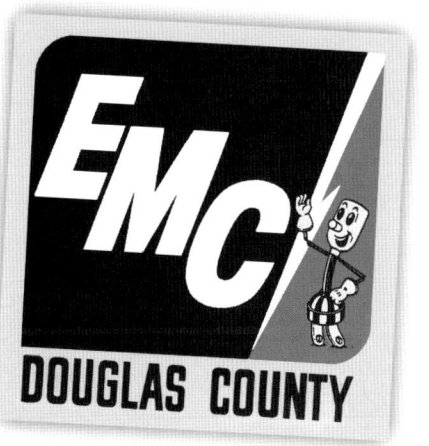

Another logo soon followed that mirrored other electric cooperatives.

chapter three

1950s: Pushing Ahead

The 1950s saw the rapid growth of the cooperative continuing at an even faster pace than the previous years. Beginning in 1950, the co-op voted to comply with the Fair Labor Standards Act and to join the statewide membership of co-ops. In 1951, there were 567 new members added during the first seven months of the year. At the August 1951 Annual Meeting, all members had received registration cards by mail with their account numbers so that they could participate in door prize drawings. The prizes were small electrical appliances, a washing machine, and a refrigerator.

The co-op had a full agenda in 1952. In this year, the EMC signed a contract with the Bell Telephone Company for the use of the cooperative's poles to bring telephone service to rural areas. Power use was discussed in May 1952 and a plan proposed to determine wiring needs and deficiencies, and a remedy for these, with the help of Douglas County EMC. During a five-month period, more than 350 new members were added. For better service a third wire was added, changing from two-wire to three-wire service. There had been an increase of about 10 percent in kilowatt-hour consumption per consumer. By September 1952, delinquent accounts had either been collected or were in the process of collection. Douglas County EMC was also investing in bonds, paying off loans, and pushing ahead with construction and improvements.

Between December 1952 and December 1953, plans were made to build extensions and increase service drops to serve about 480 members during this twelve-month period. The EMC workforce was increased, new loans undertaken, and construction pushed forward. In the first six months of 1953, 453 new members were added to the

Jessie Davis and Glen Camp at the EMC building.

cooperative alone. Member participation in the running of the EMC was also growing. At the EMC's first Annual Meeting in 1937, fewer than 100 members were in attendance. At the 1953 Annual Meeting, that number had risen to 624, jamming a local high school's auditorium. Although this represented only a small part of the organization, interest was growing, and members were becoming conscious of their roles. During the last six months of 1953, Douglas County EMC received 436 applications for membership. Surplus funds were invested in bonds and the organization began to feel solid and also grateful for sincere and conscientious directors and employees.

In 1954, applications for membership were still coming in faster than they could be serviced. Improvements were constantly being made on lines as well as office management. By April of that year, funds were available to make recommended improvements. The capacities of the Dallas and Douglasville substations were increased. In 1954, it had become apparent that the monumental growth of the cooperative both in members and energy use had outgrown the system. To correct this, new substations were installed in Douglas County near Mt. Carmel School and in Cobb County north of Austell. The substation at Palmetto had to be moved to Rivertown Road.

Mary Alice Mayfield, GreyStone Retiree

Retirees Judy Cooper (left) and Mary Alice Mayfield

Mary Alice Mayfield joined Douglas County EMC in 1950 as a cashier. Over the years, she also held positions as a capital credits clerk and a secretary to manager Josiah Abercrombie. In the 1950s, Mayfield says, they answered the phone as the "REA." There were five inside employees when she started, and she remembers the office had four desks behind the front counter and a little billing room.

Although it was all done by hand, Annual Meeting registration went smoothly at Douglas County High School at early meetings.

Chapter Three   25

Members of the board of directors reported the cooperative's current financial stance to members.

Many types of entertainment brought members to the meetings, including hearing the Douglas County High School band.

Douglas County EMC Annual Meetings were some of the most popular events to attend in Douglas County.

Before the year ended, 249 more members joined the co-op. W. A. Foster, the co-op's attorney, had the honor of being elected Superior Court Judge and resigned from his position. Robert Noland was elected to replace Foster.

The year 1955 started out busier than ever. In February, bids were opened for switching and regulator structures to be located in Dallas, Douglasville, and Mt. Carmel. New bookkeeping methods were adopted, and employees were paid every two weeks. New members were admitted each month until in May 1955, 117 new members a month seemed a normal figure. That same year the co-op sponsored a beauty contest for sixteen-, seventeen-, and eighteen-year-old girls who lived on co-op lines.

From June 1954 to June 1955, at least fifty-three miles of conversion had been completed for system improvements, rephasing, and heaving up lines from all substations. New regulators and switching materials were added to connect power lines to four substations. In April 1955, W. B. Dodson was elected by the board to fill the unexpired term of H. K. Goode. Between August and December 1955, 394 new members were added; thus ended another five years of achievement.

In early 1956, Douglas County EMC decided it would be more economical to purchase its own bulldozer for right-of-way clearing and cleanup. For the first four months of 1956, a total of 408 members were added. Efficiency was improved by the purchase of a new billing machine and addressograph machine. The corporation now had nine trucks of various types, a company car, and a caterpillar bulldozer. The cooperative was well on its way to high achievements.

On July 1, 1956, Douglas County EMC completed 50,000 hours of work without a lost-time accident and was rewarded an REA Safety Plaque. This perhaps was the start of the EMC's first safety program. It took the EMC more than a year of accident-free working days to achieve this record. Safety was and still is a top priority at the co-op.

During the last few months, Director J. H. Abercrombie had been on leave due to illness. For a temporary period, Huey Gibson was in charge. In September 1956, H. V. Latzelfelner, Rural Electrification Administration special field representative, and Georgia field representative Ralph Harper met with directors to study the co-op's business management and to assist where needed. At a meeting in 1956, the board of directors created a power consultant position to the board. J. H. Abercrombie, who served as manager of the cooperative for nearly twenty years, filled the position. Abercrombie was made available to the board of directors to give expert advice on power supply, power distribution, and other related matters.

Entertainment was sometimes groups or individuals who competed against others to win the best performance.

Douglas County EMC began talent and beauty contests for the lovely ladies in the counties it served.

The luckiest winner at this year's Annual Meeting took home an electric stove and was fortunate to meet Walter Harrison, known as the "father" of rural electrification. DCEMC Manager Josiah Abercrombie congratulates the winner.

Times were hard, and co-op members were happy to win a gallon of motor oil, aluminum foil, a tire, or a kitchen clock.

Teaching young ladies how to be home economists helped Douglas County EMC reach out to serve the community.

It was an exciting moment when a winner's name was called.

Teaching women to use the electric appliances they were purchasing became a role that co-ops took on. Employee Nancy Brown and Douglas County Extension Agent Joann Douglas taught cooking, craft, and decorating classes.

By October 1956, night service men were hired to maintain continuous service. One lineman and one crewman/ground man worked from 5 p.m. to 8 a.m. for a period of five days each. A proposed progressive wage schedule was adopted so the cooperative could operate smoothly and without animosity. To make it convenient for members to pay bills after hours, a night depository box was installed on the outside wall of the building. To better communicate with members, the first issue of *The Messenger*, a monthly newsletter, was published and mailed to all Douglas County EMC members in October 1956. The EMC continued to work to provide its members great value, and in November 1956, the directors decided on a sales promotion program and employed a home economist.

As first steps in 1957, the cooperative purchased a truck and began negotiations for more property. Norma Garner was appointed electrification advisor for Douglas County EMC during this year. In this position she gave housewives tips on the proper use of household appliances and labor-saving devices in the home and on the farm. Garner also began a series of cooking schools. The kitchen was equipped and used for demonstrations.

To provide more benefits for members, in June of 1957 the EMC, in partnership with numerous electric appliance dealers, gave members the opportunity to buy an electric range or electric water heater without having to pay extra for installation. With the purchase of an electric range or an electric water heater the co-op offered members

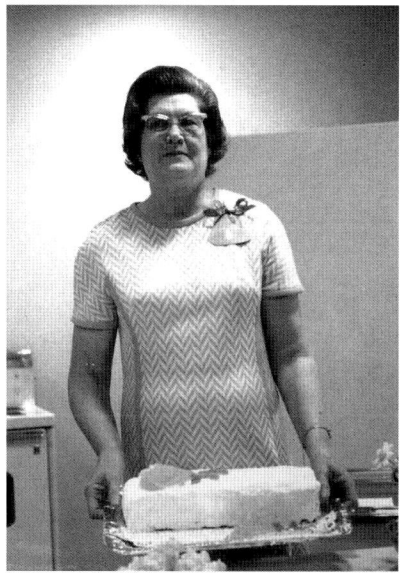

Annice Hallman baked a beautiful cake.

Tim Clower, who began at the co-op as a meter reader and rose up the ranks to become president/CEO, helped out with the Future Farmers of America.

Chapter Three **29**

Douglas County EMC's Annice Hallman took a group of ladies to a conference to help them expand their culinary and homemaking talents.

1957 EMC crew members (front row, l-r) Glenn Weathington, Doyle Ploof, Rudolph Eaves, Tim Clower, Dean Ogle, Gene Martin, Charlie Thompson, Dean Reese, and Willie B. Cox. (Back row, l-r) Wilbur Hensley, Bob Rice, Jack Hamrick, Loyd Lee, Harold Davis, Clyde Rice, Fred Webb, Roy Lawler, Robert Hunter, Whit Rice, and Charles Rutland.

free wiring and installation during the months of June, July, and August. This service saved members thirty to sixty dollars.

At the August 1957 Annual Meeting, State Representative A. A. Fowler Jr. (Sonny, son of previous Douglas County EMC Director A. A. Fowler Sr.) read a resolution to renew the co-op's charter for a period of thirty-five years. It was also reported to members that they received a reduction in rates during the past year. At the election of officers after the meeting, W. R. Thomas was elected board president of the co-op for the next year.

During the next few months, sales promotions were featured. Homeowners were given the benefit of outside security lighting, installed at a special rate. In October 1957, the board of directors approved the co-op to offer security lighting to members. The first security light installed by the cooperative was placed in front of Mt. Carmel Elementary School in that same month. Members could secure one of these lights for their yards and outbuildings for a flat fee of three dollars per month. The security lights were installed at no cost by the co-op. The lights came on automatically at dark and turned off automatically at daylight.

In addition to enhancing its services to members, the EMC worked on the development of its workers. Members of the outside crew attended Hot Line School in Americus, Georgia, and several of the office force attended special bookkeeping classes.

In September 1957, H. H. Gibson resigned as manager, and Ralph Harper was appointed acting manager of the co-op. A few meetings later, Harper was permanently elected manager. Harper moved into the community and took over his duties on November 3, 1957.

During this month, Douglas County EMC was again saddened by the death of another pioneer who worked hard to help organize Douglas County EMC. Director J. H. Abercrombie died after a long illness and a wonderful record of service.

By March 1958, office management and the distribution system were running smoothly—the co-op was pleased with the progress of the past year. The co-op civic spirit was felt everywhere. Donations to help FFA students, 4-H Clubs, little league teams, summer camps, and many other organizations were creating good public relations. In April 1958, the co-op joined the Georgia State Chamber of Commerce, and on April 17, 1958, the co-op began to host district meetings each week until a meeting was held at each district within the cooperative's service area. The purpose of the meetings was to inform members on how the co-op operates and Douglas County EMC's plans to improve services. It also gave the EMC the opportunity to speak with members one on one.

Elmer Hudson in front of the eighth substation built by Douglas County EMC. GreyStone now has thirty-three substations.

The twenty-first Douglas County EMC Annual Meeting was held on August 6, 1958, at Douglas County High School's auditorium. Coca-Colas were served after the main speaker. Local talent entertained and competed for prizes. The meeting was as much fun as a county fair. By September 1958, Douglas County EMC applied for another loan to build 180 miles of line, to serve 1,100 members, and for increased service capacity, system improvements, and engineering.

In February 1959, Manager Harper reported that co-op notes for that period had been paid in full and in advance. He also reported that co-op activities had increased as well as the yearly margin. During 1959, Harry Littler of ZIV-TV met with the directors concerning sponsoring a television program for co-ops in the Atlanta area. A contract was made with Southern Engineering Company for system improvements as discussed in the recent system study.

During this year another loyal director, R. O. Boatright, died. F. M. Boatright, son of the late R. O. Boatright, was elected to fill the unexpired term of his father.

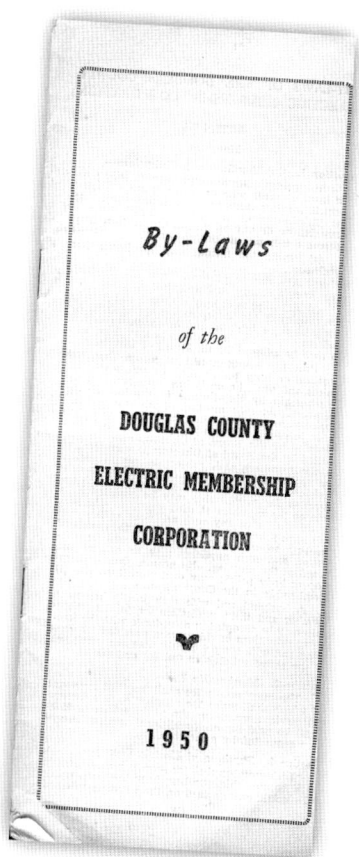

Bylaws of the Douglas County Electric Membership Corporation of 1950.

Chapter Three

chapter four

1960s: Continued Growth

At the beginning of the 1960s, the EMC had more than 8,000 members. In only twenty-five short years, the cooperative had grown from the darkened rural areas outside of Atlanta to a large electric cooperative providing its members with reliable electricity, easing the difficulties of their rural existence. As the co-op grew, so did the membership fee. On July 16, 1960, Douglas County EMC's membership fee increased from five dollars to ten dollars.

On the twenty-fifth birthday of REA, Manager Harper reported, "Douglas County EMC is stronger financially than ever before, and still growing in every respect." In July of 1963, the cooperative moved into its new building in Douglasville at 4040 Bankhead Highway. Over the years, the headquarters received several additions. Between 1964 and 1967, many improvements were made to create a more efficient operation. Data processing equipment was installed, a new accounting machine was purchased, a new policy manual was adopted, and the board approved the organization of an "Employees Welfare Association." In April 1966, Charles Overman joined Douglas County EMC as general manager.

The workforce of Douglas County EMC in 1961. (Front row, l-r) Doug Gober, Elmer Hudson, Bill Moody, Doyle Ploof, Billy Head, D. H. Kirby, Jimmy Jackson, James Campbell, Lamar Howard, Robert Hunter, Clyde Rice, Richard Holloway, Glenn Camp, and Mike Ledbetter. (Back row, l-r) Ralph Lawler, Ted Campbell, Charlie Thompson, Roy Lawler, Don Anderson, Jeanette Scroggins, Annice Hallman, Rachel Holloway, Mildred Dukes, Linda Keaton, Lois Meadows, Linda Woods, Mary Alice Mayfield, Ralph Harper, general manager, Estelle Abercrombie, Ruby Shadix, Fred Webb, Peanut Hall, Bob Rice, Charles Rutland, and Loyd Lee.

The new building was very much needed as the co-op continued to grow, with no signs of a slowdown in the number of services being added.

Land was purchased at 4040 Bankhead Highway, and soon co-op employees moved into their new home, more centrally located in the service area.

In 1967, the co-op began to participate in the Rural Electric Youth Tour, a program coordinated by the National Rural Electric Cooperative Association (NRECA) that brought high school students to Washington, D.C., each year to learn about electric cooperatives, American history, and U.S. government. The co-op offered all-expense-paid trips to the nation's capital to two winners of an essay contest. The program was open to fifteen-, sixteen-, and seventeen-year-olds. Contest entrants were required to submit a five-hundred-word essay on the subject "What local ownership of rural electric cooperatives means to our community." GreyStone delegates joined winners of similar contests sponsored by other Georgia EMCs.

A major ice storm that occurred January 9, 1968, set the co-op back $100,000 in power line damage. The EMC received help from co-ops in Florida, South Carolina, North Carolina, and southern Georgia. Much was learned from the storm. First and foremost was the implementation of a right-of-way program to clear the right-of-way throughout the system every four years.

By 1968, 95 percent of all new service built by Douglas County EMC was for residential units. Commercial and industrial development was needed to achieve the proper growth balance for the EMC; thus the co-op entered the field of community and area development. The co-op anticipated offering assistance to local governments, chambers of commerce, and others interested in the growth of west Georgia. The Douglas County EMC member service

GreyStone began to participate in the national Washington Youth Tour program in the 1960s, sending young people to the nation's capital to learn about their government up close and personal.

Chapter Four

Larry Miller (left), mapping engineer, and assistant Randy Crofts.

Douglas County EMC's employees and board members had seen firsthand the tremendous advantages that came from having electricity in members' homes and businesses. Living better just came naturally with the advent of electricity.

The mapping department completed a new inventory mapping system in the late 1960s.

Board members continued the goal of the cooperative, to supply reliable electricity to members at the lowest rates possible. The cooperative's only reason for existence is to serve members.

From time to time, the ladies of the co-op chose to wear uniforms as they assisted members.

Ladies continued to learn more and better ways to use their electric appliances in the home. Cake decorating and baking classes were popular at the co-op.

Sylvia Wilburn supervised the Paulding-Bartow district office.

department held its first series of quarterly dinner meetings for builders and contractors at the cooperative's auditorium. The first meeting was held February 20, 1968.

With the growth of the co-op's distribution system, on April 29, 1968, Larry Miller, Douglas County EMC mapping engineer, began the initial field surveys for a new inventory mapping system. As of May 10, 1969, Miller and his assistant, Randy Crofts, had mapped approximately 1,200 miles of line of Douglas County EMC's 1,600 miles of line distribution system. The crew was expected to complete mapping the system in early January 1970.

In March of 1969, the co-op announced the launch of a new concept in Georgia rural electrification—member service committees. These committees guided members through programs that enabled them to help improve the utility service, representation, and recreation facilities in their communities.

A good example of one of these programs was the development of a credit union. Douglas County EMC was instrumental in working with Georgia EMC in forming a statewide federal credit union for employees with the first meeting held at the EMC. That year Douglas County EMC also partnered with EMCs all over the nation in joining the Cooperative Finance Corporation, an organization that loaned necessary capital, in addition to the loans provided through the REA, to member-owned electric systems. To better serve members, branch offices were opened in Fulton and Paulding Counties in the late 1960s.

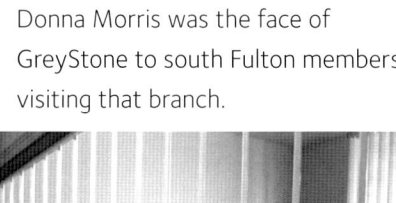

Donna Morris was the face of GreyStone to south Fulton members visiting that branch.

Chapter Four

chapter five

1970s: Overcoming Challenges

As in previous years, Douglas County EMC increased its facilities, efficiency, and service. During this decade, Douglas County EMC held its first Service Awards Banquet. A delicious meal and good entertainment set the stage as the board of directors and management extended their congratulations and thanks to employees for their years of service.

In May of 1970, Douglas County EMC signed a contract with the Industrial Development Division (IDD) of Georgia Tech to provide a program for economic development, community development, and technical assistance in its service area. In making the announcement, Douglas County EMC General Manager Charles L. Overman said, "It is our hope that through this industrial development contract we will be able to provide a program that leads toward comprehensive development of the counties in which the EMC serves. Hopefully, it will provide for county leadership, information, and answers concerning problems which heretofore have delayed comprehensive economic and community development efforts."

A major change in the meter reading and consumer billing program was implemented in 1971. The co-op began reading electric service meters in October of that year. Three meter readers were employed to overcome the inadequacies of the previous member meter-reading

The co-op held the first Service Awards Banquet in 1970 to honor employees for their years of service. Employees were photographed in groups according to years of service.

process. The meter-reading program was one of the co-op's most important achievements, bringing more efficiency to the EMC operation. Prior to the program, only 20 percent of the system's meters were read and this figure was made up primarily of public buildings, schools, churches, and commercial establishments. In conjunction with this new program, Douglas County EMC changed from a one- to three-cycle billing. This also increased efficiency by reducing the number of accounts billed during a given time period. The chances of error decreased both in reading meters and in processing bills. It also freed consumer accounts employees for other jobs. A second shift was also added to the EMC's service department operation. This made immediate service available day and night, Monday through Saturday.

Charles Overman resigned, and Charlie Thompson took over the reigns as general manager. (l-r) Charlie Thompson, Charles Overman, board chairman C. B. Peek, and board member George Hembree.

The Douglas County EMC board of directors announced the appointment of C. W. Thompson as acting general manager to fill the position vacated August 31, 1970, by the resignation of Charles L. Overman. In January 1971, the board announced their appointment of Thompson as general manager.

The co-op's dedication to quality member service was soon recognized. At the twenty-ninth National Rural Electric Cooperative Association's Annual Meeting held February 12–18 in Dallas, Texas, Douglas County EMC was awarded first place for planning and

The co-op was expanded to include a warehouse, additional office space, and renovation of the current building.

Chapter Five   **37**

The renovated building offered a drive-through night deposit.

administering the most outstanding overall member service program among cooperatives exceeding 4,500 members.

Due to the co-op's increased cost of power brought about by Georgia Power Company's 47 percent wholesale rate increase, in March 1972 the EMC found it necessary to increase rates for the first time in its history. The rates went into effect in members' April electric bills. At the close of 1972, more than 22,000 accounts were being served, with approximately 2,000 miles of line in operation.

The second major ice storm disaster in the history of the EMC occurred January 7, 1973. At the peak of the storm, 87 percent, or more than 19,000 of the EMC's 22,000 members, were without

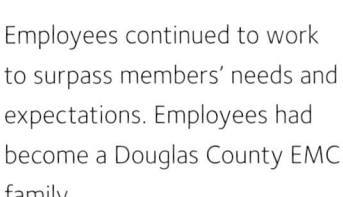

Employees continued to work to surpass members' needs and expectations. Employees had become a Douglas County EMC family.

Douglas County EMC's parade floats were always watched for in the parades—and often parade winners.

power. Remedial measures taken as a result of the experience of the 1960 catastrophe lessened the length of the outage; however, the EMC suffered approximately $214,000 in damages.

The Georgia Territorial Electric Service Act, providing for the orderly furnishing of retail electric service to the consumers of the state of Georgia, was signed into law on March 29, 1973. This legislation was critical to the EMCs of Georgia because it stopped the destructive encroachment of Georgia Power Company and municipal systems into the rural areas of the state.

In August of 1973, authority was given to Georgia Electric Membership Corporation for power supply planning and procurement. This action was the beginning of the formation of Oglethorpe Power Corporation, the power supply cooperative, of which Douglas County EMC became a member in August 1974. Also in November 1973, the sale of the old headquarters building located in Douglasville was consummated with the City of Douglasville and became the Douglasville City Hall. The corporate bylaws were revised and approved by the membership at the Annual Meeting in October of 1974, and Tim B. Clower was appointed general manager in November.

The death of line crew foreman Harvey Clark saddened Douglas County EMC. Harvey Leon Clark of Douglasville died September 25, 1974, of complications resulting from burns he suffered in a work-related accident in a substation the previous day. Clark was twenty-four at the time of his death. He came to work at Douglas County

Tim Clower was named manager of the co-op in 1974. He received expert assistance from Jerri Carruth, who would later become human resources manager.

Chapter Five

Data processing became the trend of the business world. (l-r) Corky Craven, Evelyn Hood.

Keeping up with the massive numbers of transactions as the cooperative grew from a rural area to a metropolitan one brought many challenges. (l-r) Frances Weeks, Lorraine Wilkie.

EMC on June 23, 1969, and due to his maturity and professionalism, quickly moved up the ranks.

In January 1975, the EMC expanded use of its existing computer system to perform bookkeeping functions that previously were done using a manual accounting machine. Through computer use, posting to general ledger and other functions enabled accounting personnel to do more extensive work in the areas of analysis, financial projections, and reports, as well as compiling the information for data processing. Then-accounting supervisor Mary Alice Mayfield said, "turning to data processing methods of accounting is an advancement for the EMC. In this way we are able to efficiently use our computer while keeping abreast of modern accounting methods."

With the price of wholesale electricity increasing in the 1970s, the co-op introduced members to load management. Load management helped to levelize future rates by balancing energy loads demanded within the EMC's system and lowering peak power requirements. In August 1975, Douglas County EMC offered a secondary lightning arrestor on the exterior of members' homes upon request. These arrestors, offered at the co-op's cost of $6.18, were installed at the EMC's convenience. The arrestors were designed primarily to help control minor power surges which could burn out light bulbs and damage electronic equipment.

Another disaster struck on September 23, 1975. This time, it was Hurricane Eloise, which cost the EMC $43,000. An estimated 55 percent of members were without power due to the hurricane. Working long, grueling hours in bad weather, crews were able to restore power within seventy-two hours to all members except a few isolated consumers who had to have electrical repairs made to their homes before power could be restored. Most of the damage was caused by high winds and trees falling across lines; twenty power poles were broken.

Ever since the oil embargo of 1973, cheap energy was a thing of the past. As the co-op faced rising wholesale power costs and a down economy, 1974 and 1975 were challenging years for Douglas County

EMC. Through the co-op's partnership in Oglethorpe Power Corporation, the co-op sought to ensure that an adequate future supply of reasonably priced power would be available to members. Joining this cooperative partnership was just one aspect of the co-op's plan for the future. The EMC urged members to help control power costs through load management. The concept of load management was that the EMC's wholesale power costs were based on its peak demand. Douglas County EMC's energy alert program, Operation Turn-Off (later called Watt Wiser), encouraged members to reduce consumption during peak periods in order to reduce the EMC's system overall peak load, or at least to keep the peak from increasing. The EMC gave members helpful tips and advice on how to conserve energy throughout its publications.

The property adjoining the headquarters on the west was purchased in November 1977 to expand the EMC's headquarters. On August 28, 1978, the board of directors gave the general manager the authority to proceed with the building program, which comprised new warehouse facilities, additional office space, and renovation of the present office structure. The new facilities were occupied in 1981. On December 17, 1979, the Paulding branch office was moved from downtown Dallas to Paulding Plaza.

Earlier in the 1970s, the EMC experienced annual growth rates as high as 12.73 percent. During 1979, the cooperative passed the 30,000 meter mark, yet had a more modest member growth rate of 4.73 percent. This moderate rate allowed the EMC an opportunity to prepare for the future.

The entire cooperative mourned the passing of Harvey Clark, who died from burns he received the previous day while working in a substation. He was twenty-four.

Safety and training for employees is always one of the highest priorities at Douglas County EMC.

Chapter Five

chapter six

1980s: Reaching New Heights

Forecasts indicated an upturn in the nation's economy by mid-1981. With this upturn, the EMC experienced higher growth rates and increased demand for electricity. Douglas County EMC's electrical distribution system included more than 2,400 miles of lines in 1980. Foreseeing a steady number of rate increases, and recognizing the significance of peak demand on electric rates, the cooperative worked to help members help themselves through several programs. The Demand Control Program announced in June 1981 was carried out through the use of a positive control load management system that minimized peak demand and kept rate increases to a minimum.

On May 17, 1981, the EMC held an open house of its headquarters in Douglasville, on Bankhead Highway. EMC members had the opportunity to tour the recently expanded main office

State Senator Nathan Dean welcomes Douglas County EMC members including employee Robert R. Hunter to the Open House.

building and learn more about the operations of the cooperative. After touring the building, guests met in the auditorium to enjoy refreshments and fellowship. In April 1981, the co-op rates increased, forced by a 9.3 percent wholesale power cost increase.

While 1983 was a fairly routine year for the operation of Douglas County EMC, the economic upswing brought about significant growth in requests for residential electric service. The requests for service to new homes increased dramatically from 884 in 1982 to 1,508 in 1983. An additional $3 million was required for distribution plan additions and sixty-six miles of additional distribution line to extend service to new homes and businesses. On March 24, 1983, a snowstorm caused considerable damage and power outages. The labor and materials required to repair the snowstorm damage totaled more than $58,000. Oglethorpe Power Corporation (OPC) imposed a wholesale rate increase effective December 28, 1983, which in turn necessitated an increase in retail rates that began January 10, 1984.

As the hands of time turned from 1982 to 1983, the cooperative turned a new leaf in operating techniques. In early January of 1983, the EMC completed installation of the basic equipment needed in thirteen substations for the new Supervisory Control and Data Acquisition (SCADA) system. This new operating philosophy recognized the arrival of automation in the utility industry through the increased need for ready information. The SCADA system was a computer and telecommunications network that electronically controlled various field operations and could retrieve data from the EMC's thirteen substations and report the status of station voltages, loads, and certain pieces of equipment. This allowed EMC personnel to reduce the time required to restore power after an outage and increase safety precautions for EMC employees when working on energized lines as well as prevent potential power outages. In 1983, Douglas County EMC's kilowatt-hour (kWh) sales exceeded 500 million for the first time in the history of the cooperative. The 1983 Georgia Territorial Peak Demand of 12,527 megawatts (MW) was set August 22. Douglas County EMC's contribution to the statewide peak was 127,190 kilowatts, a 28.8 percent increase over 1982's peak demand. The Demand

A new SCADA system helped the cooperative to be more informed about outages more quickly.

The Good Cents Home Program encouraged members to conserve energy.

Chapter Six 43

Douglas County EMC board members and President/CEO Tim Clower visited with Congressman Newt Gingrich about co-op concerns in Washington.

Control Program shaved 6,205 kilowatts from the EMC's contribution to the peak for a savings of more than $400,000.

Douglas County EMC announced its Good Cents Home Program in March 1985. The EMC joined the program in an effort to help members achieve the greatest energy savings possible in their new homes. This innovative program provided incentives to builders who constructed energy-efficient homes equipped with the latest energy saving features and high efficiency electric heat pumps. In this year, Douglas County EMC awarded its first Good Cents Home Certificate. The certificate was awarded to Mr. and Mrs. Steve Quigley of Fairburn, upon the completion of their Good Cents

Meter reading becomes easier with the new Datacap system used by readers like Carla Geer.

The Paulding office on Legion Road nears completion.

Home. In addition to the certificate, the Quigleys were presented a check for $475 for having their house built under the program.

For members who owned existing homes, the EMC assisted through the Energy Resources Conservation (ERC) Loan Program. This innovative program provided low interest loans to members who needed home energy savings improvements. The 5 percent interest loans were offered in amounts up to $3,500. Programs such as the Good Cents Home Program and ERC Loan Program showed the close involvement and commitment the EMC has with its members.

Some of the members of the management leadership team meet to discuss building a new headquarters building at the present site. (l-r) Tim Clower, Lynwood James, Harvey Ritch, and Charles Procter.

In May of 1986, the EMC implemented a retail rate increase of approximately 5 percent. This rate increase was necessary because of a wholesale rate increase imposed on the cooperative by Oglethorpe Power Corporation, the EMC's wholesale power supplier. However, the EMC held some of the lowest electric rates in the state. Due to 1983's record-breaking heat wave and the wholesale rate hike, the cooperative needed an additional $2 million in revenue in 1984. Without the benefits of the Demand Control Program, it would have been necessary for the EMC to raise the rates even higher. The Demand Control Program resulted in savings of more than $400,000 in wholesale power costs. At the end of 1983, 13,673 demand control switches were installed on members' central air conditioners, water heaters, and heat pumps.

Board members and others join in breaking ground for the new headquarters building.

On August 24, 1986, Douglas County EMC recorded its fiftieth year of supplying electric service to its members. May 5 and 6 of 1986, rural electric cooperatives from across the nation rallied together on Capitol Hill for their annual Legislative Conference to collectively make their voices heard as they fought for the survival of the Rural Electrification Administration (REA). Douglas County EMC was among thirty-four Georgia EMCs represented in Washington, D.C. The cooperative was represented at the conference by board president C. Billy Peek, vice president C. Worth McClure, and general manager

Joe Wallace installs underground wire.

Building begins on a new headquarters building. It will be complete in late 1990.

Tim Clower. At the conference, EMC officials met with Congressman Newt Gingrich from the Sixth District and also with Senators Sam Nunn and Mack Mattingly to express their concern over rural electrification budget proposals. The EMCs collectively urged legislative support to halt the Reagan administration's proposal to sell federal hydroelectric projects. Douglas County EMC received approximately 8 percent of its wholesale power from these federal hydroelectric projects. A sale of these projects to private utilities would have translated into an increased amount that the EMC had to pay for its wholesale power supply and, ultimately, increased rates for members.

Douglas County EMC underwent a major internal reorganization in 1987 to improve efficiency and provide better service to members. In June of 1987, Douglas County EMC began to read meters with hand-held microcomputers called Datacaps. The Datacaps gave the cooperative access to the latest technology in the electric utility industry at the time. The Datacaps replaced the cumbersome meter reading sheets that had to be completed by hand. This system immediately started to pay for itself through reduced mistakes on readings and speeding up the billing process. Digital mapping was another great improvement for the co-op. In 1988, digital mapping was added to the EMC's drafting area. This modern concept sped up mapping procedures and allowed for greater efficiency.

In the spring of this year, Douglas County EMC announced its association with the Salvation Army's Project Share Program. This unique program allowed EMC members to donate up to five dollars per month, to be added onto their electric bills, to help local people in need. Project Share was a program established by the Salvation Army in cooperation with the electric utilities of Georgia to help address the energy-related need for those on low and fixed incomes, elderly, disabled, and other identified needy individuals. These emergency needs included such items as food, shelter, clothes, and help with utility bills.

On May 22, 1988, the cooperative celebrated the completion of its new Paulding district office located at the corner of Legion Road and Dallas-Acworth Highway in Dallas, Georgia.

October 4, 1988, marked a watershed date for the cooperative. On that day, members at the Annual Meeting voted overwhelmingly (399 to 169) to change the name of the cooperative from Douglas County Electric Membership Corporation (EMC) to GreyStone Power Corporation. The previous name was somewhat restrictive in accurately describing the cooperative's eight-county service area. Because large deposits of granite, "grey stone," are found in all eight counties the cooperative serves, and mined in seven of those counties, GreyStone Power Corporation was selected as the new name of the cooperative. A group consisting of members of the Douglas County EMC management team and board of directors selected the name after countless hours of research and debate.

To keep up with the increased workload, GreyStone's number of employees almost doubled by 1988. At the end of 1988, GreyStone had a total of 47,050 meters, with more than 3,145 miles of power lines. During 1989 GreyStone implemented a 4.5 percent rate increase but still had some of the lowest electric rates at that time. GreyStone's new three-hundred-foot radio tower, stationed at the cooperative's main office, provided the co-op with better radio communication, improving efficiency over its eight-county service territory. It was time to build a new headquarters facility in order to continue the level of service that our members deserved.

Jeff Camp (left) and Steve Findley are trained on aspects of the newest meters.

Well-wishers gathered to celebrate the opening of the new Paulding district office building on Legion Road in Dallas.

chapter seven

1990s: Preparing for the Future

The 1990s were a new page in the cooperative's history, packed with many goals and challenges. In 1990, the cooperative opened a new headquarters building and added 1,600 new members to its system, for a total of 50,762, and eighty-four miles of new power lines. Because of this continued growth, GreyStone's workforce also expanded to keep the quality of service to members at the desired level. By the close of 1990, GreyStone employed 220 people. With the completion of a new Mt. Carmel substation in Douglas County, which replaced an older substation that had to be removed because of the widening of Highway 92, another modern facility was added to GreyStone's system for more than 3,300 miles of distribution lines. GreyStone also instituted an electric water heater program in 1990 that provided rebates to homeowners who replaced their old gas water heaters with energy efficient electric water heaters. The results of this program also helped to provide a more balanced load factor.

GreyStone Power Corporation continued to evolve in 1991. In the fall of that year, GreyStone announced that a retail rate increase would go into effect in January of 1992. This was brought about in large part due to the wholesale rate increase of GreyStone's wholesale power supplier, Oglethorpe Power Corporation. Wholesale power charges made up more than 75 percent of GreyStone's costs. While the

In 1990, GreyStone opened a new headquarters building at its 4040 Bankhead Highway site.

wholesale rate increase was substantial to the cooperative, as much as feasible it was operationally absorbed to make the retail rate increase as minimal as possible.

In October of 1991, the co-op hosted its open house and Annual Meeting for the first time in its new headquarters. At the event, GreyStone's roots were showcased as a memorabilia exhibit. The exhibit served as a tangible effort to thank the many people who played a part in the co-op's tremendous success and to enshrine forever those members who began the co-op in 1936 with only eighty-three members.

Michael Craton teaches children safety around electricity—one of GreyStone's highest priorities.

The number of meters served grew to 55,041 in 1992. Now near completion, the conversion of the entire system from 7,200 to 14,400 volts continued in 1992, preparing the cooperative to meet its electricity needs. Educational programs were presented in schools by GreyStone employees with the newly acquired "Safety City," stressing safe behavior around electricity. Service continued to be rendered at two district offices in Fulton and Paulding Counties. Bartow and Carroll members enjoyed the convenience of a toll-free telephone number, direct to the cooperative. The larger, consistent loads of industrial customers improved system efficiency and helped to keep rates affordable. In 1992, commercial, industrial, and residential sales totaling 14,372,564 kilowatt-hours were added to GreyStone's member base.

Begun in 1985, the Good Cents Home Program saw 299 units certified in 1992. Under residential marketing manager Phil Landress, the program provided homes of maximum comfort and efficiency for their owners and conserved energy at the same time. A new program called NightBright allowed members to install security lights with no installation charge and three months of free service. The overwhelming response and very important benefit of making members feel safer in their homes led GreyStone to consider running the program on an annual basis. The W.H.A.M. (Water Heating And Maintenance) program, also born in 1992, allowed members to switch out old water heaters for new ones, or switch out gas water

They called it "the blizzard of the century" when it hit in March 1993, putting 20,000 GreyStone members in the dark and the cold.

GreyStone linemen braved the elements to turn heat and lights back on for members.

heating systems for electric models with a monetary rebate from the cooperative. In addition, members could choose to pay a low fee on their monthly bill to receive water heater maintenance. GreyStone was also one of the first to explore and use the benefits of ground source heat pumps, and in this year, Barton Place in Carroll County became the first subdivision in the metropolitan area to exclusively offer underground geothermal heat pumps. GreyStone also improved services to members by offering an automatic bank draft payment plan.

As GreyStone Power entered 1993, many challenges were ahead. The first came in the form of a March blizzard with storm conditions unlike any the co-op faced before: the storm of the century. Although 20,000 GreyStone members were without power at the height of the storm, half had power restored by the first night, with the remainder back on line by the second evening. As a result of the storm, GreyStone incurred more than $300,000 worth of damage; however, the damage was minimized by the excellent right-of-way work performed prior to the storm by GreyStone.

With a sigh of relief, GreyStone headed into spring and summer. However, relief was short-lived, as Georgia experienced one of the longest heat waves on record. For sixty-nine summer days, temperatures registered at least 90 degrees. The three months of summer in 1993 saw a 24 percent increase of energy use over the same period in 1992, while new members grew by 4.2 percent during those same months. GreyStone's supplier, Oglethorpe Power Corporation, broke every generating record of its history during the scorching summer months. In the heat of the battle, GreyStone consumer service representatives did everything possible to help members offset high energy bills. Because of the crisis, GreyStone management delayed a planned rate increase until 1994.

In addition to the on-line performance, GreyStone's team practiced preventive tactics to help keep rates down. These tactics included a concentrated effort to install as many demand control switches on members' heat pumps and air conditioners as possible, resulting in 24,000 total switches. The new switches helped combat rate increases due to the co-op's ability to cycle off compressors by radio signal at critical moments. GreyStone was able to lower the peak demand for electricity, which determined the rate, thereby saving more than $1,022,000 in wholesale power costs.

At the end of 1993, the cooperative that began on a shoestring budget with eighty-three members in 1936 had 56,766 meters and 3,569 miles of power lines. At the close of 1993, GreyStone had streamlined its staff to 205 employees in accordance with recommendations of a reorganizational study, down from 222 at the end of 1992. At this time GreyStone was the second largest cooperative in the state of Georgia in number of meters per mile, just behind Cobb EMC.

Congressman Phil Gingrey takes time from his busy schedule to meet with GreyStone directors.

Communicating with national and local officials like Congressman David Scott is a top priority for GreyStone.

Chapter Seven

Co-op employees were stunned to lose lineman Travis Allen to the Hurricane Opal restoration effort. "He was doing what he wanted to do," his mom, Belle Allen, said. Foremen often asked Allen to show rookies the ropes due to his prowess on the lines.

Veteran GreyStone employee Tracy Worthan died following a car accident that occurred while on duty, less than a month after Allen's death. GreyStone employees grieved the loss for his family and the GreyStone family.

In May of 1994, electric cooperatives including GreyStone Power representatives gathered in Washington, D.C. Co-op officials met with Senators Sam Nunn and Paul Coverdell along with Congressmen Buddy Darden and John Lewis to lobby for refinancing of Oglethorpe Power's federal loans, and other issues which would ultimately help all EMC members.

Packing winds up to eighty miles per hour, Hurricane Opal left almost 96 percent of GreyStone's 60,000 members without power during the early morning hours of October 5, 1995. According to President/CEO Tim Clower, the devastating storm caused more damage to the system than any he had witnessed to date. "Because of the enormous number of outages, many members had difficulty reaching us by phone. We had employees manning the phones around the clock during the outage, but the system was simply overwhelmed because of the tens of thousands of calls trying to get through," said Clower. Broken poles numbered eighty-two, and the total cost of damage to GreyStone's system was approximately $1.5 million.

Tragedy struck October 11 when GreyStone employee Travis Allen sustained severe injuries while restoring power to one of the last homes affected by Hurricane Opal.

Supervisory Control and Data Acquisition employee Hal Murphy and GreyStone linemen and engineers worked to restore the outages that covered the GreyStone system.

Downed trees and tree limbs made restoration of power particularly difficult following Hurricane Opal's destruction.

Employees David New, Lynwood James, and Tommy Paine cook meals for the linemen and deliver food to them on the job so the restoration effort can continue.

Allen, one of GreyStone's most universally respected apprentice linemen, died October 25, 1995, after a valiant fifteen-day struggle for life in Grady's Burn Unit. Travis came to GreyStone May 11, 1987, and had earned the rank of apprentice lineman IV. He had completed the TVPPA Lineman's Apprentice Program with excellent scores, and often was an example to other GreyStone employees during safety training for his climbing and lineman abilities. A blood drive was held on November 3, 1995, in honor of Travis at GreyStone at the family's request, continuing his desire to help others, as evidenced by his volunteer work as a Shriner.

Less than a month after Allen's death, GreyStone lost another employee in the line of duty. Service technician Tracy Worthan, age sixty, died November 15, 1995, as a result of injuries sustained in a work-related traffic accident on November 13. Worthan came to work

Billy Peek (left) and Tim Clower go over a little co-op business before the board meeting. Tim is to retire after forty years of service to the cooperative.

GreyStone employees (l-r) Bill Sharpton, Gary Miller, and Lisa Jones look at outage statistics in SCADA.

for GreyStone on August 24, 1972, after retiring from the military. A twenty-three-year employee, he had worked as a meter reader, senior meter reader, and section manager of meter reading, and was a service technician at the time of his death.

At GreyStone's request, GEMC Federal Credit Union (GEMC FCU) opened its GreyStone offices in June of 1996, making GreyStone the only electric cooperative in the state, and one of the first in the nation, to offer credit union membership to its members. GEMC FCU welcomed many new members on November 1, 1996,

GEMC Federal Credit Union opened at Douglasville and Dallas in June of 1996.

Gary Miller becomes president/CEO of GreyStone Power Corporation in January 1999.

at the credit union's open house held at both the Douglasville and Dallas GreyStone locations. There were 248 guests registered. All GreyStone members and their families were eligible to join. To become a member of the credit union, GreyStone members had to complete a membership application and return it with five dollars or more to open a credit union savings account. GEMC FCU offered checking and savings accounts, as well as a wide range of financial products and services.

In the summer of 1997, GreyStone established its first interactive website, www.greystonepower.com. The website featured information about GreyStone programs and services available to members. Additional benefits to members were instituted including the convenience of credit card payments for electric bills. GEMC FCU's VISA card offered members a 1 percent rebate off all new charges on their electric bills. In 1997, members also received another great perk. A residential retail rate reduction went into effect April 1, 1997, averaging approximately 8 percent for most GreyStone members. The reduction was a result of combined efforts by GreyStone and Oglethorpe Power Corporation to maintain costs.

GreyStone introduced three new programs for members at its October 1997 Annual Meeting that provided value-added services at very cost-competitive prices. The new programs were Power Paging, a paging service providing local service for as low as $6.95 monthly;

The first volunteer GreyStone Power Foundation Inc. board used members' rounded-up dollars to soften life's blows for members needing help through local charitable organizations.

When not restoring power or signing up new GreyStone members, employees are hard at work on other causes, like beating cancer. GreyStone employees support many causes, with the three main drives being Relay for Life, March of Dimes, and United Way.

Chapter Seven   55

GreyStone and other cooperatives reach out to countries around the world to help them learn how to bring electricity to their homelands. GreyStone VP Tim Williams hosts a delegation from Indonesia and is thanked by the head of the group.

SurgeMaster, a surge and lightning protection program for homes and businesses for only $5.95 per month; and GreyStone Security, a security installation and monitoring service with monthly monitoring for $12.95.

Nineteen ninety-seven was a banner year, as rates actually decreased for members.

In August 1998, GreyStone offered members weather radios for $78 each. The seven-channel weather radio was programmed by code to alert members of weather specific to their county.

That winter, GreyStone signed a marketing alliance with SCANA Energy to market natural gas to its members. GreyStone President/CEO Tim Clower believed the new alliance with SCANA Energy would allow GreyStone to offer members natural gas at competitive rates. At the time, SCANA Energy was also partners with three other EMCs including Cobb, Snapping Shoals, and Central Georgia EMCs.

In November 1998, GreyStone Power Corporation began an innovative program that provided financial assistance to worthy organizations and individuals in its service area. Through Operation Round Up, each participating cooperative member made a small monthly contribution by voluntarily signing up to round their electric bill up to the next dollar. Together, those small amounts made a big difference in people's lives and a tremendous impact on GreyStone's communities.

GreyStone Security began protecting members' homes and businesses in October of 1997.

As 1998 came to a close, GreyStone's torch was passed to a new president/CEO as Tim Clower ended his forty-three-year tenure at the cooperative. Clower was the president/CEO of the cooperative for twenty-five years. GreyStone Power's board of directors selected Gary A. Miller as the new president/CEO to guide the cooperative as it entered the new millennium.

Georgia's population grew by 26 percent in the 1990s, and GreyStone was ready to meet the increased demand. In the summer of 1999, GreyStone and other participating Georgia electric cooperatives added a 217-megawatt combustion turbine at the Smarr Energy Facility near Forsyth in Monroe County. GreyStone's portion of this facility was 15 megawatts.

In the summer of 2000, four combustion turbine units generating 492 megawatts were added at Sewell Creek near Cedartown in Polk County. These units were used to handle peak demand. GreyStone's portion at Sewell Creek was 19 megawatts. GreyStone had 65,000 members in 1999 and had unprecedented growth in the retail arena.

Working in a substation is one of the most dangerous tasks a lineman faces.

chapter eight

2000s: Embracing a New Millennium

Much was said about the impending millennium. GreyStone left no stone unturned in preparing for the calendar's entrance into the year 2000 and its corresponding computer concerns. Having met its goal of 100 percent Y2K (Year 2000) compliance in September 1999, GreyStone did not expect any problems occurring as the clock struck 12 a.m. on January 1, 2000. The co-op had visited every scenario that could occur and prepared contingency plans for overcoming them.

As growth skyrocketed, GreyStone's board of directors and management looked at service in every area to find ways to improve it. The co-op added new line crews to build and maintain power lines more effectively and to restore power outages faster. The use of twenty-one large peaking generators at several strategic substations helped the co-op to keep costs down for members. In 2000, for the first time, members could pay their bill, sign up for service, perform energy audits, or get information for other services or products on the GreyStone website. The cooperative announced in 2000 there would be no rate increase in 2001. This marked the ninth consecutive year without a rate increase. In 2000, GreyStone ranked among the top three cooperatives in service in the nation, according to figures compiled by the National Rural Electric Cooperative Association (NRECA).

In April 2001, Oglethorpe Power announced plans to build six units of a $280 million peak power facility in Talbot County near the Harris-Muscogee county line. At the time, Talbot Energy Facility's total capacity was 648 megawatts with GreyStone's share totaling 133 megawatts.

At the end of 2001, GreyStone finalized a ten-year deal, set to start in 2005, with the Southern Company for a 50-megawatt supply from

the company's coal block. In 2001, GreyStone members paid less for electricity per kilowatt-hour than they did in 1991. The cooperative purchased a new telephone system to enable employees to handle members' requests in a more efficient and timely manner. This interactive voice response (IVR) system gave members additional options for service; they could speak with a member services representative or carry out their business through keypad options.

A new telephone outage reporting system was also implemented to allow quicker response to outage-related calls. In addition, locating devices were placed in strategic areas throughout the system to automatically report disruptions in service to GreyStone's control center. This enabled the cooperative to have crews on the scene faster than ever. New software helped to streamline the administrative response process to widespread outages and allowed the co-op to have a better handle on how best to respond at all times. GreyStone also extended member services representatives' working hours to allow members to speak by phone with a "live" person until 7 p.m. Monday through Friday.

In 2002, GreyStone Power continued to make significant advances in its ability to serve members, both technologically and in the cooperative's infrastructure. A new parking deck that provided

GreyStone was stunned like the rest of the nation on September 11, 2001, when terrorists destroyed the World Trade Center, killing thousands of Americans. The theme of the October Annual Meeting was "United We Stand."

Chapter Eight

two hundred parking spaces improved access and convenience for members. The deck, which opened at the cooperative's 2002 Annual Meeting, was built in anticipation of the eventual expansion of Bankhead Highway to four lanes.

A new cooperative, Green Power EMC, encouraged members to sign up for power generated through environmentally friendlier, alternative energy sources. The cooperative began taking subscriptions on Earth Day, April 22, 2002, and by the end of the year reported the purchase of nearly six hundred Green Power EMC blocks. GreyStone was one of sixteen co-ops in Georgia to incorporate Green Power EMC, the first renewable energy program begun by utilities in the state of Georgia.

The installation and operation of a new Supervisory Control and Data Acquisition (SCADA) system allowed GreyStone dispatchers to better supervise substations and other devices in the field, pinpointing actual or potential problems and reducing response times during outages. Another service and infrastructure improvement was GreyStone's new 29,000-square-foot warehouse addition. The warehouse helped the cooperative to keep pace with its rapid growth. The facility comprised two floors and 1,800 square feet of office space.

During 2003, GreyStone added more than two hundred miles of power lines and more than 3,100 meters. Rates remained stable and the announcement of no rate increase for 2004 marked the twelfth consecutive year without a retail rate increase. As in years before, GreyStone continued to work to keep service interruptions

Congressman Mack Collins was the keynote speaker as the birth of Green Power EMC was announced at the Southface Energy Institute in Atlanta in 2002.

to a minimum. For the five years through 2001, GreyStone's average outage time per member was 1.23 hours, compared with the Georgia average of 3.76 hours. GreyStone reached out to members in a new way in 2003 and implemented a guest services booth at Arbor Place Mall in Douglasville. GreyStone's guest services booth gave members easier access to GreyStone information and increased the co-op's visibility in the community.

Since 1996, GreyStone Power membership increased by 58 percent, assets grew by almost $143 million, and demand for electricity surged 77 percent. But, most notably, members' equity increased by $38 million. In 2004, GreyStone Power's growth reflected the economic health of its communities and the daily dedication of employees. GreyStone's growth also strengthened its ability to provide improved services and greater benefits to members. During 2004, the co-op took a giant step in improving member services with the creation of a stand-alone call center for faster, personalized service.

GreyStone's investment in equipment, an outage management system, and the professionalism of GreyStone linemen paid off as the co-op's service area was hit not once, but twice by the remnants of destructive hurricanes. The aftermath of Hurricanes Frances and Ivan left up to 17,000 members without power; the co-op's automated outage reporting line received almost 24,000 reports. GreyStone crews worked around the clock to remove fallen trees and repair power lines. Within two days, only 1 percent of the outages remained. With power restored to members, GreyStone linemen volunteered to help restore power in Florida. One six-man crew from GreyStone saved the life of a Florida man whose generators had poisoned his store.

From left, Aaron Parrott, Matt Williams, Patrick LeCroy, Eddie Elrod, and Matt Gilbert (front center) were honored with Life Saving Awards from both GreyStone and Georgia EMC for saving a life in Florida during hurricane relief efforts. (Not pictured: Justin Galloway)

Six GreyStone Power linemen received the Georgia EMC Life Saving Award in 2004 for their lifesaving efforts while working on storm damage in Florida. The six volunteers were helping restore power to the ten thousand members of Escambia River Electric Cooperative following Hurricane Ivan when they responded to cries for help from police officers. A husband and wife had collapsed in their fish market near Pensacola, but sheriff's deputies could not enter the building because of burglar bars on the windows. The GreyStone crew used their line truck's large winch to remove the bars and gain entry. The woman died at the scene, but her husband was revived at the hospital. An investigation indicated that the couple suffered carbon monoxide poisoning while trying to refrigerate fish using three generators inside the building.

The installation and operation of a new Supervisory Control and Data Acquisition system enabled faster restoration of power.

During 2004, GreyStone grew to more than 5,443 miles of power lines and more than 94,684 meters. Rates remained stable in 2004, and a new power supply agreement with Progress Energy Ventures Inc. shielded GreyStone members from sharply increasing energy costs through 2010.

Skyrocketing energy prices forced a modest increase in retail power rates in 2005 for the first time in more than a decade. Higher natural gas and coal prices translated into higher electric power costs, but GreyStone Power was able to keep rates among the lowest in the country. The turmoil in neighboring states caused by Hurricane Katrina tested the courage, ingenuity, and energy of GreyStone linemen who volunteered to help restore power.

Energy was on everyone's mind during this year. Gasoline pumps and front-page headlines screamed crisis in the price and availability of energy. GreyStone wasn't immune. Sharply increasing natural gas and coal prices translated into much higher electric power costs on the open market than the year earlier; however, GreyStone Power rates remained among the lowest in the state.

From left, Jim Hunter, Tommy Farmer, and Gary Miller go over plans for the new Dallas office as it comes up out of the ground.

GreyStone continued to grow at a healthy 7 percent rate during 2005. A new 10,000-square-foot Dallas district office building opened in December 2005 to serve the growing needs of GreyStone Power members in Paulding County. The facility featured enhanced security and telecommunications services and multiple drive-through service windows for faster and more convenient service to members. The building also had the capacity to act as a backup to GreyStone's headquarters in Douglas County in an emergency. The office serves members on GreyStone Power Boulevard, near the intersection of Highway 278 and Cadillac Parkway. This year GreyStone also responded to members' service needs by expanding call center operation to twenty-four hours a day, seven days a week.

Finally, the death of Board Chairman C. Billy Peek saddened the entire GreyStone family, who held him in high esteem. Peek served as a member of the board of directors for twenty-five

A new 10,000-square-foot Dallas district office building opened in December 2005.

Member services representative Debbie Suttles greets members with a smile at the Dallas district office.

years and twenty years as chairman. Previously, his father, C. B. Peek, was a director for twenty-eight years. Calvin Earwood, a member of the GreyStone Power Board of Directors for twenty-eight years, was elected chairman, succeeding C. Billy Peek.

As GreyStone's membership grew, so did the need for financial services. In July 2006, GEMC Federal Credit Union celebrated a decade of service to the GreyStone community. At the time, the credit union had more than doubled in size, serving 11,000 members, since opening its membership to the GreyStone community. The credit union offered many additional services for members, such as a car buying service, child safety identification program, scholarships, and auto insurance. All GreyStone Power members and their families were eligible to join GEMC FCU to take advantage of the great rates, outstanding service, and numerous financial options.

While providing reliable energy is at the core of GreyStone Power's mission, the cooperative formed an EMC Security partnership in October 2006. GreyStone joined Jackson Electric Membership Corporation (EMC) and Walton EMC to offer security and fire protection, along with other technologies such as advanced wiring, central vacuum, home theater, whole house audio, video, and a wide range of commercial and industrial products and services. GreyStone's contract with SCANA Energy was set to expire May 29, 2008, but essentially ended in January 2006. In October 2006, GreyStone entered into a new natural gas contract with Gas South, joined by other Georgia EMCs to help thousands of EMC members save on their natural gas. The co-op signed a marketing agreement with Gas South that initially allowed the co-op to only market its new natural gas partnership to new members who did not have SCANA until February 2008.

In early 2006, a photovoltaic system began to brighten Hiram High School in Paulding County thanks to a unique partnership between GreyStone Power, Green Power EMC, and the Paulding

County School System. The joint venture provided a powerful hands-on learning experience about solar energy for the school's students while generating valuable research information needed for creating sustainable, inexpensive solar technology. Still in place today, the system converts sunlight into energy through silicon alloys called photovoltaic cells. In addition to the meter that records how much electricity is produced, the system included a data collection program that monitors weather conditions and direct current and alternating current power production. This data is then sent to an interactive website that can be used to monitor, evaluate, and compare the performance of all photovoltaic systems installed in schools across Georgia.

Balancing members' needs for energy with concern for the environment and balancing income and expenses in order to continue affordable rates were major goals for the co-op in 2007. In this year, the cooperative purchased one Toyota Prius vehicle and leased two others to wisely use fuel. GreyStone used a 20 percent alternative fuel mixed with diesel fuel to power the fleet of sixty-four different vehicles, including trucks, tractors, and trenchers. With this change, GreyStone used 18,000 fewer gallons of fossil fuel than it would have otherwise. The same employees who operated those vehicles also recycled parts used in building and maintaining GreyStone's power delivery system. Four large

GEMC Federal Credit Union celebrated its tenth anniversary of serving GreyStone members in July 2006.

Chapter Eight

GreyStone has returned $43 million in capital credits to members over the years.

bins on each work dock provided a place for linemen to recycle these items after a long day working on the lines.

Office employees recycled everything from computer parts to paper. In 2006, a shredding service collected and recycled eleven tons of paper from the Douglasville and Dallas offices. GreyStone Power also replaced every incandescent bulb in its offices with compact fluorescents.

Three residential marketing representatives worked with GreyStone members to help them use energy more efficiently as well. One of the most comprehensive ways was through free home energy audits that helped members save energy and money. Many families benefited from this free service in 2007.

In 2007, GreyStone joined Touchstone Energy Cooperative, a nationwide network of electric co-ops that are dedicated to bringing added value and benefit to members. Touchstone Energy is the brand name by which cooperatives identify and connect themselves with that alliance.

Being a member of GreyStone Power has its privileges, as members saw in fall of 2007 when they received a Co-op Connections Card. Touchstone Energy's exclusive membership discount card offers cooperative members thousands of local and national deals on a wide range of services, including prescriptions, oil changes, spa treatments, restaurants, day care, chiropractic services, and jewelry. In 2007, more than 150 businesses, locally and across the nation, offered discounts to GreyStone Power members through the Co-op Connections Card.

Supporting area businesses by helping them use energy resources wisely is essential to economic development at

GreyStone Power. The co-op's trained professionals provided businesses with expert energy advice. Specific answers could be found through online energy analysis and energy libraries on the co-op's website, www.greystonepower.com. Professional on-site consultations and services, including walk-through energy audits and infrared thermal imaging, were available to members of GreyStone. Commercial members began receiving an online electronic newsletter, *Questline*, which offered energy saving tips.

Commercial members could also support Green Power EMC, which plugs into "green" resources such as biomass, solar, and water to generate electricity. The Inn at Serenbe was GreyStone's first commercial member to support the effort. As Green Power members, The Inn helped fund research into new ways to develop cleaner, greener energy in Georgia, like Plant Carl. The renewable energy plant was originally set to generate 20 megawatts of electricity as the first poultry litter-

GreyStone Security was purchased by EMC Security in October 2006; GreyStone joined Jackson EMC and Walton EMC in offering EMC Security to help keep members safe.

Chapter Eight

A unique partnership between GreyStone Power, Green Power, and the Paulding County School System helped teach students about solar energy.

to-energy operation in Georgia. Green Power participants also supported green power generation at several solar, landfill, and hydroelectric sites.

As of December 31, 2007, there were a total of 4,117 active customers enrolled in EMC Security services. GEMC Federal Credit Union had a total membership of 11,280 as of April 30. Of those, 4,652 or 41.2 percent were GreyStone Power members. In 2007, GreyStone had more than seventy years of experience in servicing the co-op's utility vehicles. In 1995, as the need for service grew dramatically, Fleet Services became a full-service operation for fellow EMCs, telephone companies, and other companies with utility fleets. In 2007, Fleet Services serviced and maintained GreyStone's 176 pieces of equipment and provided repairs and maintenance for forty other companies. This resulted in outside revenues of more than $430,000.

In this year, GreyStone members made their voices heard in the halls of Congress. Through the Our Energy, Our Future campaign, organized by the National Rural Electric Cooperative Association, co-op members aimed to influence lawmakers by pressing them on the climate change debate. More than 2,100 GreyStone members and employees sent letters and e-mails to their representatives, the most of any Georgia cooperative.

In many ways, 2007 and 2008 held many challenges for GreyStone. As a business, the co-op faced an economic recession, rising costs for supplies, fuel, power/capital constraints in financial markets, and an uncertain regulatory environment. In September and October of 2008, growth slowed, and the co-op's membership declined for the first time in our history. But within two months, GreyStone made up for those losses. In January through May 2009, the co-op saw positive growth, but at a rate one-tenth of previous years.

During slower periods, the cooperative turned its attention to the future, beginning new projects that would translate into advances in service and reliability. GreyStone shifted employees into planning roles and made plans for new accounting, work asset management, and customer care and billing systems. These would greatly improve service, provide members with additional information, and serve as a framework for new services.

The reduced demand for labor forced GreyStone to release 250 as-needed contractors, who worked on new construction, and led the co-op to freeze hiring. At the same time, GreyStone's call center showed a 15 to 20 percent increase in calls as members lost jobs, faced other financial hardships, or needed assistance from GreyStone's member service representatives. To no dismay, the cooperative was able to keep rates competitive. Surveys consistently ranked GreyStone as having some of the lowest rates in the state and the nation. A variety of programs and partnerships assisted members as well, including the co-op's participation in Touchstone Energy's Co-op Connections Card program. As of November 2008, GreyStone members had received more than $455,000 in prescription benefits alone and saved at 290 participating local businesses. In 2008, GreyStone also began the Cooperative Healthy Savings program to assist members with dental, vision, and hearing costs.

Members also benefited through GreyStone's continuing alliance with natural gas provider, Gas South. GreyStone members receive discounts on all standard residential Gas South rate plans. GEMC Federal Credit Union continued to offer members great rates and dividends. Increased self-service

Residential marketing representative Lisa Lonon works with a builder to help him incorporate money-saving features in the homes he is building.

Chapter Eight

In 2007, GreyStone joined Touchstone Energy Cooperative, a nationwide network of electric co-ops that are dedicated to bringing added value and benefit to members.

options, including online checking and billing, reduced the cost of doing business, as well as environmental impact.

As GreyStone moved toward a crisis in having enough power generation to meet demand, energy efficiency and conservation were more important than ever. In 2008, economic indicators and changing market dynamics led the cooperative to the decision to withdraw from its commitment with other electric cooperatives to build Plant Washington, a state-of-the-art coal plant. GreyStone planned to replace electricity generated at the proposed facility with natural gas-fired resources.

Through Green Power EMC, a partnership that now comprises thirty-eight electric cooperatives in Georgia, GreyStone continued to expand renewable energy options. In 2008, Green Power EMC was the first utility in Georgia to earn the distinguished Green-e Energy certification, which recognizes excellence in meeting rigorous environmental standards.

In February 2008, GreyStone Power celebrated the milestone of adding the co-op's 100,000th member, Adrian Williams of Fairburn. He and his family expressed excitement about once again being served by an electric cooperative, GreyStone Power. To celebrate the co-op's milestone of growth for the community, GreyStone Power friends and partners gave Williams and his family housewarming gifts valued at more than $3,000. "I thought it was a joke—I kept laughing and thought someone was playing a trick on me, but it's real," said Adrian.

As another first, members cast their board election ballots in a new way in 2009. The board of directors "vote by mail" process enabled members to more easily have a voice in who represents them at GreyStone Power.

GreyStone Power earned safety accreditation from the National Rural Electric Cooperative Association (NRECA) with flying colors in November of 2008. In addition, the cooperative completed the 2008 year with no lost-time accidents.

As GreyStone's communities struggled to recover from a tough economy, in 2009, GreyStone Power focused on what has worked best for members of the cooperative since day one in 1936—working together to save. While providing more than just electric service, GreyStone worked to save members money and energy. "Helping the people we serve save money begins with controlling our own costs and keeping electric service bills some of the most affordable in the state and nation," said GreyStone President/CEO Gary Miller. Inside the co-op, the hiring freeze continued. Retiring employees were not replaced, and existing employees took on additional responsibilities.

Even though the cooperative added 1,040 new electricity services in 2008, the number was down from a high of 6,200 around 2007. GreyStone's maximum demand for electricity was also down to 651

Being a member of GreyStone Power has its privileges. "Just try the discount Co-op Connections Card," says GreyStone employee Ginny Bailey.

Chapter Eight 71

GreyStone's Fleet Services became a full-service utility vehicle repair option for outside utility companies in 1995, and in 2007 serviced and maintained GreyStone's 176 pieces of equipment and provided maintenance and repair for forty other companies.

megawatts from a high in 2007 of 891 megawatts. Energy sales were down; yet the electricity delivery system needed to deliver power still had to be maintained.

GreyStone increased efficiency by setting up rate values for construction, maintenance, service-related tasks, and right-of-way clearing. Rather than using only on-call crews in the evenings and early morning hours, GreyStone assigned regularly scheduled crews to be used as needed by on-call crews.

The cooperative energized new substations in the Cliftondale and Annewakee areas and made improvements to its existing substations. During outages, GreyStone began to restore power to sections of circuits if the co-op was not able to repair the entire circuit. This enabled GreyStone to restore power to unaffected areas served by those circuits even faster.

In 2009, GreyStone offered more online services at www.greystonepower.com. In addition to paying their bills online, members could now submit more than thirty-five online forms related to their electric service. This made service more convenient while keeping staff costs more manageable.

Helping members manage their energy use wisely through energy efficiency education became an even higher priority in 2009.

In 2002, President/CEO Gary Miller was named Citizen of the Year by the Douglas County Chamber of Commerce.

Adrian Williams of Fairburn and his family became GreyStone's 100,000th member household in February 2008.

GreyStone Power practiced what it preached by running thermal imaging studies on its heating and cooling systems and made energy-efficient upgrades to save money and energy.

In July 2009, Burnell Redding celebrated forty years serving members on the board of directors. Redding was president of the board from 1981 until 1985 and also served many years as secretary-treasurer.

GreyStone replaced inefficient lighting with energy-efficient bulbs and installed occupancy sensors so lights turn off automatically in unoccupied rooms. The cooperative adjusted thermostats, as well. GreyStone's goal was to decrease overall energy consumption by 5 percent in 2010.

Wise energy use for the future has also been a major concern for the co-op. GreyStone's continuing Our Energy, Our Future campaign let Congress know the need for affordable, reliable energy solutions for years to come. By the end of 2009, more than 2,700 members had sent almost 13,500 messages as a result of GreyStone's efforts.

Safety and loss control director Blake Pendley and president/CEO Gary Miller with the NRECA plaque awarding safety accreditation.

Other clean energy efforts include $1.2 billion invested in new emission control additions by GreyStone's wholesale electricity supplier, Oglethorpe Power. The controls significantly reduce known pollutants.

The "Together We Save" website and "You Have the Power to Save" communication materials helped members use energy more wisely in 2009, saving money and helping ensure GreyStone will have the energy needed for the future. The immediate savings on bills were evident. The new Sun Rays Power program that was launched in 2008 also helped by providing as much as 70 to 80 percent of a member's annual hot water needs at a fraction of the cost of heating water with gas or electricity. Both state and federal tax credits helped offset the cost. GreyStone Power also offered a $500 rebate to those participating.

New technology helped the cooperative monitor its electric distribution system as well. In 2007, GreyStone began the task of replacing approximately 116,000 meters with technologically advanced meters. "Smart" meters revealed patterns of electricity use and enabled GreyStone to expand its services by offering prepaid electric service to members. Prepaid service would end the need for

In September 2009, monumental rains brought flooding to metro Atlanta, with Douglas, Paulding, and Cobb Counties some of the hardest areas hit. In some cases, it took using boats to repair the damaged lines.

Training linemen for any eventuality in case of an accident is an ongoing process at GreyStone.

deposits for those participating. Reliability and cost-savings were considerations in GreyStone's decision to add these meters. The information they provided would eventually help decrease the co-op's response time during outages. They also allowed GreyStone to read meters remotely, saving labor and expenses.

In September 2009, three days of rain brought monumental floods to metro Atlanta, with Douglas, Paulding, and Cobb Counties some of the hardest hit areas. The storm washed away homes, roads, vehicles, and their occupants, and left almost 37,700 GreyStone members without electricity. Linemen worked around the clock in

In 2004, GreyStone's board of directors voted to name its lineman training facility the Charlie W. Thompson Training Facility in honor of former manager Charlie W. Thompson.

sixteen-hour rotations and contracted eleven crews to restore power more quickly. Crews drove boats into flooded areas to make needed repairs. Two of GreyStone's linemen rescued a stranded young man who was fleeing the rising water on the top of his car. One of GreyStone's directors also waded through chest-deep water to save a neighbor in her home. In all, four GreyStone employees and one director were honored by Georgia Electric Membership Corporation (GEMC) with a "Life Saving" Award in 2010. The GEMC Life Saving Award was established to recognize EMC employees across the state whose courageous and skillful actions helped someone in grave and immediate danger.

Initial estimates showed the co-op experienced more than $537,615 in damages to the electricity delivery system. The Federal Emergency Management Agency reimbursed GreyStone for many of the expenses, so those costs did not have to be passed along to members through higher rates. GreyStone Power was able to assist the community with repair costs as a part of flood relief efforts by donating $365,000 to the Douglas County United Way, Paulding County United Way, Douglas County School System, Paulding County School System, and directly to local schools.

GreyStone Power was founded in 1936 on the concept that by working together, we all could save money and have a much-needed service: electricity. Today, GreyStone is still working for the benefit of its membership.

For the first time, GreyStone's 2010 Annual Meeting of Members featured an exclusive Together We Save Energy Efficiency Expo, showcasing the value of energy efficiency. The Expo was abuzz with members learning how to become more energy efficient and wise energy managers in their homes and businesses. Members who registered at the Annual Meeting received a free Home Energy Savings Starter Kit. Several specialized businesses educated members about the benefits of conservation and demonstrated valuable energy-efficient products and services. GEMC Federal Credit Union offered low-interest energy-efficiency loans and the Expo featured an "Ask the Energy Specialist" Q&A booth led by GreyStone's very own energy experts.

On Earth Day, Green Power EMC agreed to a long-term purchase agreement for 17 megawatts (MW) of energy produced by Multitrade Rabun Gap, the first biomass plant in Georgia. The plant began to produce energy January 28, 2010, and the participating EMCs hosted a ribbon cutting at the plant on Earth Day, April 22.

On May 12, 2010, GreyStone Power approved a five-year Power Purchase and Scheduling Services Agreement with Morgan Stanley Capital Group valued at more than $600 million. Morgan Stanley Group is a subsidiary of Morgan Stanley engaged in wholesale sales and purchases of electricity throughout the United States, including Georgia. The agreement provided that

GreyStone Power Foundation Inc. board members celebrate with winners of Operation Round Up scholarships.

From left, Tony Brown, Patrick LeCroy, Josh Jones, and Tim Costner won first place at the Georgia Lineman's Rodeo, first place in the EMC division at the 27th Annual International Lineman's Rodeo, and second-place Overall World Champion Journeyman team at the international event.

Morgan Stanley would schedule the energy from GreyStone's resources or provide power from the market, whichever was more economical, to serve all of GreyStone's load. This agreement allowed GreyStone to adapt to changing legal, public policy, and regulatory requirements, to purchase renewable and alternative energy, and implement demand response, net metering, and other new technologies.

In an effort to better communicate with members, GreyStone Power embraced the fast-paced world of social networking by launching a Facebook page, Twitter, and an electronic newsletter (E-Link). With more than 250 million active users logged in on any given day, Facebook has become a necessity to keep up with today's generation. The new GreyStone Facebook page and tweets engage members with valuable energy saving tips, information about other programs, upcoming events, photos, and more.

Another great milestone occurred in the fall of 2010. GreyStone Power linemen Patrick LeCroy, Josh Jones, Tony Brown, and coach Tim Costner captured first place in the EMC division at the International Lineman's Rodeo in Kansas City and second place overall in competition with all divisions. The journeyman team competed against 155 teams. They also won third place in one journeyman event and fourth in the other. The winning journeyman team holds the title for overall winners in the 2009 and 2010 Georgia Lineman's Rodeo. Prior to competing

GreyStone employees create award-winning floats for the Fourth of July Douglasville parade.

in the International Rodeo, the team was presented with specially cast state champion rings by the Georgia Lineman's Rodeo Association. Brown, Costner, Jones, and LeCroy were the first linemen ever to earn the state champion rings.

The International and Georgia Lineman's Rodeos were both established to recognize the profession of line work. These rodeos bring together linemen to compete in events based on traditional lineman tasks. It also serves as a showcase for the education, skill, and safety training of electric linemen and helps keep linemen safe on the job.

In the fall of 2010, GreyStone Power Corporation returned a total of $3.5 million to members across portions of the eight counties served by GreyStone. When members sign up to receive electric service from GreyStone, they become a member-owner of the cooperative. While investor-owned utilities return a portion of profits back to their investors, electric co-ops operate on a not-for-profit basis and periodically return capital credit margins based on how much you paid the co-op for electricity during a specified time period. GreyStone has returned more than $47.5 million to members over the years.

GreyStone outside personnel keep the lights on, the meters running correctly, equipment at-the-ready, and the right-of-way cleared to prevent outages.

Chapter Eight

chapter nine

Working Toward a Bright Future

Who We Are Today

The history of GreyStone Power Corporation has been one of continuous growth. As a cooperative owned by the people it serves, GreyStone Power has grown from providing reliable, affordable electricity to offering ancillary services through Gas South, GEMC Federal Credit Union, and EMC Security. From solutions for keeping costs down to helping with energy efficiency, today the co-op is finding new ways to save energy and money by working together inside GreyStone Power and outside with the people we serve.

Our mission is to provide reliable and cost-competitive electric and related services that position the cooperative as the utility of choice. And we work to achieve our mission by daily living our vision to create and sustain value for our members through demonstrated leadership in the energy distribution business.

Cooperatives worldwide generally operate using the same principles adopted in 1995 by the International Cooperative Alliance. These principles, as with many co-ops, are part of GreyStone's statement of identity.

Voluntary and Open Membership

Cooperatives are voluntary organizations, open to all persons able to use their services and willing to accept the responsibilities of membership.

Democratic Member Control

Cooperatives are democratic organizations controlled by their members, who actively participate in setting policies and making decisions.

When members see a GreyStone truck coming down the road to their homes or businesses, they know help is on the way.

Members' Economic Participation

Members contribute equitably to and democratically control the capital of their cooperative.

Autonomy and Independence

Cooperatives are autonomous, self-help organizations controlled by their members.

Education, Training, and Information

Cooperatives provide education and training for their members, elected representatives, managers, and employees so they can contribute effectively to the development of their cooperatives.

Cooperation Among Cooperatives

Cooperatives serve their members most effectively and strengthen the cooperative movement by working together.

Concern for Community

While focusing on member needs, cooperatives work for the sustainable development of their communities.

GreyStone was named one of Atlanta's Best Places to Work by the *Atlanta Business Chronicle* in 2010.

Today, GreyStone Power serves more than 103,500 members in portions of Bartow, Carroll, Cobb, Coweta, Douglas, Fayette, Fulton, and Paulding Counties. Out of those counties served, Douglas County leads with 40,403 services, followed by Paulding County with 39,793 services and south Fulton with 17,770. GreyStone is among the largest of Georgia's forty-two EMCs and one of the largest of more than 900 EMCs in America. According to figures, GreyStone ranks the fourth largest EMC in Georgia in total members served with 103,500 members and fourth largest in kilowatt-hours sales. For 2010, 2,732,382,190 kilowatt-hours were sold by the cooperative. GreyStone is the sixteenth largest cooperative in the nation in number of members served.

In addition to our main office at 4040 Bankhead Highway in Douglasville, Georgia, GreyStone also has a full-service district office at 120 GreyStone Power Boulevard in Dallas, Georgia. Office hours are from 8 a.m. to 5 p.m. weekdays. Employees are on duty twenty-four

GreyStone's award-winning public relations department brings home national awards every year for their communications with members.

Chapter Nine

Giving back is synonymous with being a GreyStone employee.

Boys & Girls Club members say thank you to GreyStone for continuing support.

Helping our men and women serving in the military is one of the things GreyStone does best. The cooperative was awarded the Patriot Award in 2011 for its ongoing support of the National Guard and Reserve.

hours a day and line crews are always on call for emergencies. Call-in service is also available twenty-four hours a day.

Several major organizations with large electric power needs have selected GreyStone Power as their provider of choice over other providers throughout the years. Some of them include:

American Red Cross	PepsiCo
Colgate Palmolive	Publix
Douglas County Schools	Quaker Oats Company
Fulton County Schools	Target
Ingles	Turano Baking Company
Kroger	Unilever
Nioxin	United Natural Foods
Paulding County Schools	Wal-Mart

GreyStone is an equal opportunity employer with more than 265 employees. Directors hire a president/CEO to carry out the daily operation of the cooperative. GreyStone hires professional employees to design, construct, and maintain the electrical facilities, as well as employees who are responsible for billing, accounting, record-keeping functions, and providing information and technical service to members.

In 2010, the cooperative was named one of Atlanta's Best Places to Work. Of four hundred companies nominated, GreyStone placed seventeenth among the top twenty medium-sized workplaces in Atlanta. In 2011, the cooperative was named the top employer in Douglas County by those responding to a survey by the *Douglas County Sentinel*. The cooperative also received the Patriot Award in 2011 for its consistent support of the military. GreyStone received a Cooperative Innovators Award in 2008 from the Cooperative Research Network. The Douglas County Chamber of Commerce also named GreyStone "Large Business of the Year" in 2008. GreyStone employees are people of integrity: the true grit of members who began the co-op seventy-five years ago, and of the employees and members who have carried the tradition up to the present day. It's integrity that keeps any business serving customers for seventy-five years.

Membership in GreyStone Power has its privileges. In addition to providing the people we serve with reliable electricity at competitive rates, the cooperative is proud to offer members an array of valuable products and services including:

Car Solutions
Convenient Payment Options
Co-op Connections Card
Cooperative Health Savings
Customized Billing Options
Efficiency Loans
EMC Security
Energy Audits
Energy Saving Tips and Advice
Fleet Services
Gas South
GEMC Federal Credit Union
Green Power
Operation Round Up
Pet Assure
Rebates
SurgeMaster Plus

From left, GreyStone lineman Julio Villegas, National Rural Electric Cooperative Association President/CEO Glenn English, and GreyStone lineman Erik Hansek. The GreyStone linemen were honored for traveling to foreign countries to assist in bringing power. Lineman Keith Bailey has also been active in the NRECA International Foundation's rural electrification programs.

After the devastation of Katrina in New Orleans, GreyStone linemen volunteered to go and assist and saved a life.

Chapter Nine

GreyStone is a friend to education and supports school systems in many ways. From left, Sam Land, Ed Cahill, John Archer.

GreyStone sponsored and hosted the National Rural Electric Cooperative Association's community project last fall at Lithia Springs High School.

GreyStone Power continues to provide members with valuable discounts and incredible savings from local and national retail businesses participating in the Co-op Connections Card program. At the same time, businesses benefit through increased purchases by co-op members. Since its debut in fall 2007, this free member benefit card further proves that being a member of GreyStone Power has its advantages. With more than twenty-two million cards and key fobs in use across the country and nearly $1 million saved on prescription drugs alone by GreyStone members, the Co-op Connections Card is a proven success.

With the implementation of GreyStone's Customer Care and Billing (CCB) system on April 4, 2011, the cooperative offers faster,

The GreyStone Power Foundation Inc. volunteer board makes a tremendous difference in the co-op's service area.

more efficient services to members. New service applications allow members more flexibility to access their accounts through an enhanced and interactive website and improved interactive voice response (IVR) phone system. Members also received a completely redesigned, easier-to-read electric bill with added details about their electric usage. This more efficient system will ultimately save paper and save members time.

At GreyStone, safety is top priority. The co-op's Safety Excellence Program was created to increase safety awareness and to encourage employees to take an active role in safety. Since the program began, there has been a significant increase in safety awareness, and a notable decrease in hazards and accidents. In 2010, GreyStone completed over 365 days of no lost-time accidents. The co-op continues to keep safety first and works to achieve its safety goals.

Harriet the Heat Pump spread cheer at the 2010 Annual Meeting in October.

A Cooperative That Cares

Throughout its history, GreyStone has supported many local civic and charitable organizations. Today, employees from the cooperative serve as project volunteers and members of the Board of Directors for many organizations and clubs including the Cultural Arts Council of Douglasville-Douglas County, Relay for Life, March of Dimes, Boys

Lamar Crawford was the lucky winner of the truck in 2010 at the Annual Meeting.

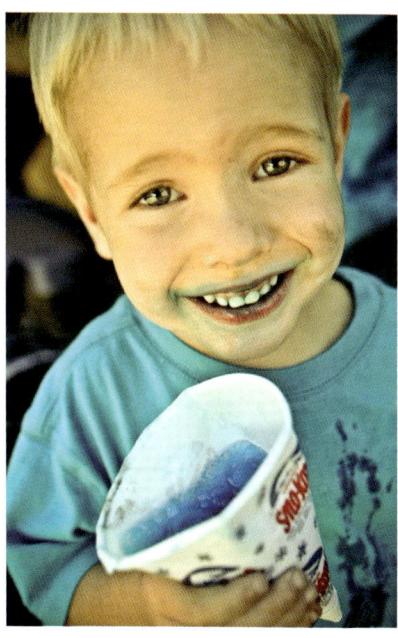

Giving back to the community is never more apparent than at GreyStone's Annual Meeting.

Chapter Nine

Preparing to welcome home the troops from Iraq, linemen Chad Voyles and Aaron Parrot place the flag on the line truck boom.

Awaiting the troops.

& Girls Club, Literacy Council, S.H.A.R.E. House, Sertoma, Rotary Club, and United Way. For multiple years, United Way in Douglas County honored GreyStone Power's employees, directors, and retirees with the Commitment to Community Award. In 2010, members of the GreyStone family donated $20,500 to United Way, helping more than 28,000 children and adults in Paulding and Douglas Counties. Through payroll deductions and donations, more than four hundred programs—like the Boys & Girls Club, Douglas Senior Services, and Shepherd's Rest Ministries—benefited.

The cooperative takes pride in giving back to the communities it serves. Among its involvement, helping to support members and their families achieve their educational goals is one of GreyStone's most successful and rewarding contributions. Annually GreyStone coordinates four unique opportunities for students; the GreyStone Power Foundation Scholarships, Walter Harrison Scholarships, Clower Scholarships, and the Washington Youth Tour. The GreyStone Power Foundation Scholarships are awarded to GreyStone Power members or their children who are seniors in high school or enrolled in college. The recipients are selected by the Foundation's Board of Directors based on academic achievement, community involvement, and need. Since 1999, more than $2.43 million has helped to support GreyStone's communities through the GreyStone Power Foundation Operation Round Up program.

Student apprenticeships/internships are other ways GreyStone supports youth development and student learning. The co-op also participates in local youth apprenticeship, mentoring, and partners in education programs. GreyStone supports the Douglas County School System's Public Education Trust Fund Inc. (PET) to help award grants and scholarships to assist students, teachers, and programs not already funded by local, state, and federal dollars. One of GreyStone's most generous contributions in 2009 was its support for the West Georgia Technical College "Invest in the Next Generation" major gifts campaign.

Together We Save

A main focus for GreyStone has been to work with members to help them better manage their energy usage. From 2008 to 2009 GreyStone featured a section in its member newsletter, the *GreyStone Report*, highlighting tips and ideas shared by members to conserve energy.

In 2009, GreyStone rolled out its Together We Save campaign, a nationwide energy-efficiency campaign designed by GreyStone and other Touchstone Energy Cooperatives to inspire members to easily save energy and money. Along with classic mediums—TV, print, radio, and billboards—the campaign has a substantial online component, www.togetherwesave.com. Members can take control of their energy costs by visiting www.togetherwesave.com, which offers valuable energy- and money-saving advice. The website demonstrates how taking simple energy-saving steps leads to real dollar savings. After visiting the site, GreyStone encourages members to share their Together We Save experience by sending in their testimonials on the actions they took to save energy and the impact made by their efforts. Selected testimonies are featured in the *GreyStone Report* to share with other members. GreyStone's

GreyStone retirees Cub Wilson (left) and Robert Hunter discover a pole that was one of the original Farmers Electrical Association poles.

Achieving member smiles is what GreyStone is all about.

Chapter Nine

The people make GreyStone what it is today—the members, the employees, working together, achieving the miracle of electricity every day.

Member Matters newsletter also features energy-efficiency articles and columns. Ask the Energy Specialist is a section in *Member Matters* that addresses issues to help members reduce their energy use to save money.

GreyStone's 2009 Annual Meeting event focused on energy efficiency and conservation through GreyStone's Together We Save campaign. The co-op hosted its first ever energy efficiency expo at the 2010 Annual Meeting. The Together We Save Energy Expo showcased the value of energy efficiency. The 2009 GreyStone Annual Report published in 2010 also focused on the Together We Save theme.

Together We Save took on an additional meaning in 2010, when four GreyStone employees and a director were honored for saving lives. Faced with extraordinary challenges and circumstances, these men and women, and many other GreyStone employees through the years, went beyond the call of duty to rescue someone in dire need of help.

GreyStone's website, www.greystonepower.com, features valuable links and tools to help members get the most out of their energy use. The co-op also promotes energy efficiency through special publications, billboards, ads, and displays. GreyStone's Facebook and Twitter pages feature valuable energy and money-saving tips. The cooperative offers residential and commercial members free energy audits to help members identify potential problems and solutions to save energy and money. GreyStone's new smart meter technology helps the co-op expand its services by offering prepaid electric service to members. As of July 2011, approximately 112,000 automated meter reading (AMI) meters have been placed, representing 92 percent of GreyStone's meters. Prepaid service helps participating members be better managers of their energy, allowing them to purchase power as they use it.

Current Challenges

Climate change concerns are prompting calls for legislative action on the federal level that would translate into much higher electricity costs for users. These actions could escalate to the point that there are those who can afford electric power in their homes and businesses and those who cannot. Power costs are also increasing as the

Linemen change out a cross arm.

From left, Josh Jones, Anita Helton, Charles Rutland, Joe Nicholson, and Scott Buchanan received statewide Life Saving Awards in 2010.

costs to produce electricity increase. And although overall growth in new services has diminished due to today's economic climate, the demand for electricity per household is increasing due to newer and increased ways to use power, especially in technology. Permits for new plants to generate electricity are routinely turned down, so the new plants that would be needed to supply enough power, should the economy turn around and growth in services increase, are not being built. The electric service arena could be headed for a perfect storm.

Also, Congress has been debating, but never passed, a comprehensive climate change bill. Into this void, the Environmental Protection Agency has stepped forward and has now fielded new regulations: the clean air transport rule, cooling water intake requirements, and the possible designation of coal ash as a hazardous material. This is of great concern for GreyStone, and the co-op continues to urge members to let Congress know of the need for affordable and reliable electricity.

Over the past seventy-five years, first as the Farmers Electrical Association, then Douglas County EMC, and today GreyStone Power, the cooperative has met the challenges before it through the support of its loyal members who are the heart of GreyStone Power. Meeting a challenge is nothing new to electric cooperatives, and drawing from the wisdom of our past experiences, we will continue to successfully serve our members.

At GreyStone, you're not a number, you're a member. Welcome to where Members Matter.

Chapter Nine

chapter ten
Co-op Leaders

GreyStone Power Corporation Managers and Years of Service

GreyStone Power Corporation's board of directors hires a president/CEO to carry out the daily operation of the cooperative. Through their leadership and GreyStone's well-trained employees, GreyStone works to provide members with reliable electric service at the lowest possible cost.

Gary A. Miller, 1999 to date

Gary Miller has been employed at GreyStone for twenty years. His leadership has helped make GreyStone Power one of the most financially stable, respected, and innovative cooperatives in the nation.

He formerly served as vice president of the financial services division. He is a graduate of North Georgia College, is a certified public accountant, and holds a law degree from Georgia State University. Before coming to GreyStone, he worked at Jackson Electric Membership Corporation in Jefferson, Georgia, and at Amicalola EMC in Jasper, Georgia.

Miller serves on and chairs numerous committees at the statewide level on behalf of GreyStone Power and Georgia's Electric Membership Corporations (EMCs). Presently, he chairs Georgia EMC's Government Relations Committee and also serves on the GRESCO board of directors. Miller was appointed to the board of CoBank in 2006 where he presently serves on the Audit Committee.

Gary A. Miller

Miller is also active in the community. He is a past chairman of the Douglas County Chamber of Commerce and the Douglas County United Way. Miller serves on the boards of directors of the Douglas County Development Authority, the Hospital Authority of Douglas County, and the WellStar Health System Board. Miller is an advisory board member of Regions Bank and a graduate of Leadership Douglas. He is a member of Central Baptist Church in Douglasville, where he also teaches Sunday school. In 2002, Miller was named Citizen of the Year by the Douglas County Chamber of Commerce for his community and business support.

Tim B. Clower, 1974 to 1999

Tim Clower came to work for Douglas County EMC in 1955 as a stock clerk. He soon advanced to lineman on a service truck. Clower's work duties were interrupted at that time by a stint in the military from September 1959 until August 1961. Upon returning, he became work order clerk and draftsman. His new title of wiring and heating specialist received at the end of 1966 was left behind as he was promoted to department manager of member services. In June of 1974, Clower was given the title of acting general manager, followed by general manager in December of 1974. In 1988 he became president/CEO, a position he held from 1988 until 1998. Two of his most prized achievements include the NRECA National Achievement Runner Up Award in 1969 and winner of First Place in Member Services in 1970. His leadership helped the cooperative to become one of the most financially secure electric cooperatives in the state and nation. Clower worked for the cooperative for more than forty-three years and served the last twenty-three years of his employment as manager or president/CEO of GreyStone Power. The Tim B. Clower Scholarship is awarded annually through GreyStone's Operation Round Up program to two deserving technical students in honor of Tim Clower and his dedication to technical education.

Tim B. Clower

Charlie W. Thompson

Charles L. Overman

Ralph L. Harper

Charlie W. Thompson, 1970 to 1974

On September 1, 1970, Douglas County EMC Board of Directors announced the appointment of Charlie W. Thompson as acting general manager to fill the position vacated August 31, 1970, by the resignation of Charles L. Overman. In January 1971, the board announced his appointment as general manager. Before being named general manager, Thompson was employed by the EMC for twenty-four years, during which time he served in several capacities. He had served as assistant general manager for five years. Thompson was instrumental in instituting training and safety programs on a statewide basis through Georgia EMC that have prevented injuries and deaths over the years. In 2004, GreyStone's board of directors voted to name its lineman training facility the Charlie W. Thompson Training Facility in honor of former manager Charlie W. Thompson.

Charles L. Overman, 1966 to 1970

Charles Overman joined Douglas County EMC as general manager in April 1966, bringing experience gained by working five years with EMCs in North Carolina, three years with the North Carolina statewide association of EMCs, and two years as district manager of another EMC in Georgia. Overman came to Douglas County EMC with many new ideas that helped the EMC grow. During his tenure, EMC operations were divided into departments, each of which had specific responsibilities. Prior to 1966, there was little decision-making below the general manager level. Douglas County EMC became the first in Georgia to computerize its operations during Overman's tenure.

Ralph L. Harper, 1957 to 1965

Ralph L. Harper became manager of Douglas County EMC in 1957. Harper came to the position after twenty years of service working as operations field representative of the Rural Electrification Administration. Harper was a native of Hagerstown, Maryland, and moved to Georgia in 1939. He was a graduate of the Milwaukee School of Engineering, where he received a Bachelor of Science in Electrical Engineering degree.

Huey H. Gibson, 1956 to 1957

In 1956, Huey H. Gibson was selected by the board of directors to succeed J. H. Abercrombie as manager. Gibson served as Douglas County EMC's manager for a short period, resigning from his position a year later in 1957.

Josiah H. Abercrombie, 1938 to 1956

Josiah H. Abercrombie was one of the five original founders of the Farmers Electrical Association in 1936. In 1938, Abercrombie and six others were elected as the co-op's first board members. Less than two years from the birth of the corporation, Abercrombie was elected manager of the EMC. At a meeting in 1956, the board of directors created a power consultant position for the board. J. H. Abercrombie, who served as manager of the cooperative for nearly twenty years, was given this appointment. Abercrombie was made available to the board of directors to give expert advice on power supply, power distribution, and other related matters.

GreyStone Board of Directors

Territory served by GreyStone is divided into nine geographical districts, and each is represented on the board of directors by a cooperative member residing in that district. Directors' terms of office are staggered to provide that three positions expire annually. Contested board elections are elected by mail-in ballots; uncontested seats are filled by voice vote at the Annual Meeting of Members held each year on the second Saturday in October.

Huey H. Gibson

Josiah H. Abercrombie

The GreyStone Power Board of Directors (l-r): Chairman Calvin Earwood, District 1: Paulding and Bartow Counties; President/CEO Gary Miller; Milton Jones, District 7: Fulton County; Charles Rutland, District 3: Douglas and Paulding Counties; Secretary-Treasurer Jennifer DeNyse, District 5: Carroll and Douglas Counties; Burnell Redding, District 4: Carroll and Douglas Counties; Vice Chairman John Walton, District 2: Paulding County; Jim Johns, District 8: Douglas County; Maribeth Wansley, District 6: Fulton, Fayette and Coweta Counties; Ed Garrard, District 9: Cobb County.

GreyStone Power Corporation Directors and Service Record

N. P. Barker 1936–1938
R. O. Boatright 1936–1959
J. S. Bomar 1936–1941
H. V. Branan 1936–1964
W. G. Johnson 1936–1938
A. A. Fowler 1936–1943
J. H. Abercrombie 1936–1938
Glenn Florence 1938–1940
J. R. Mann 1938–1962
W. H. McClendon 1938–1945
J. M. Aderhold 1938–1939
W. H. Tanner 1939–1946
C. G. Rakestraw 1940–1966
H. H. Cook 1940–1954
E. T. Evans 1940–1946
R. C. Morris Sr. 1943–1944
H. H. Jones 1945–1965
W. A. Foster Jr. 1945–1949
C. S. Keith 1946–1951
A. H. Stockmar 1946–1949
W. R. Thomas 1949–1965
W. C. Abney 1949–1956
           1964–1976
C. B. Peek 1951–1980
H. K. Goode 1954–1955
W. B. Dodson 1955–1965

E. I. Raykowski 1956–1964
F. M. Boatright 1956–1969
J. A. Keith 1962–1965
E. O. Hilderbrand 1964–1967
J. W. Pinkerton 1965–1976
James H. Wright 1965–1989
W. George Hembree 1965–1995
Cleveland B. Rakestraw 1965–1995
C. Worth McClure 1965–1993
Raymond L. Vaughn 1967–1991
Burnell Redding 1969–present
W. McSel Pearson 1976–1998
Calvin Earwood 1977–present
C. Billy Peek 1980–2005
Charles Rutland 1989–present
Fred Wallace 1991–2009
William Parks 1993–2005
Ed Haley 1995–2001
Robert Sprayberry 1995–2003
Ed Garrard 1998–present
John Walton 2001–present
Jennifer DeNyse 2004–present
Milton Jones 2005–present
Maribeth Wansley 2005–present
Jim Johns 2009–present

about the author

Zakesia (Kizzy) Howell of Douglasville, Georgia, is the former public relations and communications coordinator for GreyStone Power Corporation. She came to work for GreyStone in 2002 as a student apprentice and was promoted through progressively responsible positions in the department over her nine years of service.

As public relations and communications coordinator, Kizzy was primarily responsible for communicating with and educating internal and external audiences on cooperative policies, programs, and services. Growing in her profession, Kizzy was honored for her abilities by the National Rural Electric Cooperative Association (NRECA) through the Spotlight on Excellence Awards program for several years. Her honors included awards for design, writing, layout, and photography. Kizzy also served as a community figure on the cooperative's behalf through Leadership Douglas, S.H.A.R.E. House, and other local organizations.

Kizzy is a graduate of Douglas County High School in Douglasville and received her Bachelor of Business Administration degree from Mercer University with a focus in management, marketing, and communications.

Kizzy loves spending time with her family, especially her husband and children. She enjoys music, dancing, and watching movies.

Kizzy says she has always been fascinated with history and jumped at the opportunity to coordinate the production of GreyStone's history book. "It was a real treat putting this important historic document together. I gained a new respect for my job and GreyStone in doing so!"

GreyStone's *Lighting the Way* history book, authored by Zakesia (Kizzy) Howell and edited by Ashley Kramer and Vicki Harshbarger, is a product of Greystone's Public Relations and Communications Department.

Vicki Harshbarger
Department Manager
Ashley Kramer
Public Relations & Communications Specialist
Amanda Busby
Public Relations & Communications Specialist
Trisha McBee
Public Relations & Communications Admin. Assistant/Event Planner